Many-Body Boson Systems

Theoretical and Mathematical Physics

The series founded in 1975 and formerly (until 2005) entitled *Texts and Monographs in Physics* (TMP) publishes high-level monographs in theoretical and mathematical physics. The change of title to *Theoretical and Mathematical Physics* (TMP) signals that the series is a suitable publication platform for both the mathematical and the theoretical physicist. The wider scope of the series is reflected by the composition of the editorial board, comprising both physicists and mathematicians.

The books, written in a didactic style and containing a certain amount of elementary background material, bridge the gap between advanced textbooks and research monographs. They can thus serve as basis for advanced studies, not only for lectures and seminars at graduate level, but also for scientists entering a field of research.

For other titles published in this series, go to
www.springer.com/series/720

André F. Verbeure

Many-Body
Boson Systems

Half a Century Later

 Springer

André F. Verbeure
K.U.Leuven
Institute for Theoretical Physics
Celestijnenlaan 200 D
3001 Leuven
Belgium
andre.verbeure@fys.kuleuven.be

ISSN 1864-5879   e-ISSN 1864-5887
ISBN 978-1-4471-2612-6   ISBN 978-0-85729-109-7 (eBook)
DOI 10.1007/978-0-85729-109-7
Springer London Dordrecht Heidelberg New York

British Library Cataloguing in Publication Data
A catalogue record for this book is available from the British Library

Mathematics Subject Classification (2010): 22E70, 37A60, 46L57, 46N55, 47D06, 47N50, 81Q80, 81S05, 81T05, 82B10, 82B26, 82D50

Cover design: eStudio Calamar S.L.

Printed on acid-free paper

Springer is part of Springer Science+Business Media (www.springer.com)

To Ivonne and to our family

Preface

The writing of this book has of course been stimulated by the exciting developments in the field of Bose-Einstein Condensation (BEC) for atomic gases that have manifested since the 1995 experiments. These experiments are showing brand new features never before observed. Their theoretical analysis and understanding is however still based on the standard theory of Bose-Einstein condensation developed for space-homogeneous boson systems. Just as importantly are the recent exact results and new views on the problem of many-body physics developed during the last fifty years. Of course, many of these results have their own particular impact on the problem of BEC for boson systems. Moreover many of them seem to be only known by a small number of mathematical physicists but are less known by the larger community of physicists.

Faced with this situation, this book is conceived to be an introduction to these new concepts and results written with considerable attention toward the physical ideas behind the more technical material. Apart from the study of general and universal properties of fully interacting boson systems, numerous homogeneous boson models are explicitly treated. The applications of the presented material to systems of trapped bosons is only briefly discussed and treated as a posed problem in Chap. 4. Further study is left to the care of the interested reader.

Much of the material mentioned in the text was obtained during many years of collaborations with many colleagues and former students. Warm thanks to all of them. We feel obliged to point out one colleague in particular, Mark Fannes. Thank you, Mark, for our long standing collaborations and continuing exchanges about views, ideas, and techniques along all these years during which we constructed together much of the backbone of the present text.

2009 *Leuven*

André F. Verbeure

Contents

1

Introduction

The study of boson models and the problem of their solutions is as old as the free Bose gas model for which the celebrated phenomenon of Bose-Einstein condensation (BEC 1924) has been detected. This phenomenon puts in evidence a macroscopic, purely quantum phenomenon. Solving boson models means that we are interested in finding the ground and/or temperature states of the models, or that we are interested in deriving at least some of their properties from first quantum-mechanical principles. For a long time activities in this area belonged to the field of many-body physics, a field of high activity in theoretical physics. Green's functions, series expansions, Feynman graphs and their summations, and much of numerics are the standard technologies. The ultimate aim is, as always, to understand the physical world from the point of view of its basic laws and constituents and, therefore, to derive as many exact results as possible having a large or universal range of validity for these systems.

From the point of view of exact results in this field of physics, the work of Araki-Woods [12] greatly boosted the understanding and clear formulation of the properties of the free Bose gas and its accompanying phenomenon of Bose-Einstein condensation. Careful studies of the thermodynamic limit for the free Bose gas also inspired efforts toward finding exact results [101]. In both cases these works relate the problem of the equilibrium states of the Bose gas to a problem of representations of the canonical commutation relations (CCR). This problem is by itself as old as the early days of quantum mechanics. Indeed the basic problem associated with the foundations of quantum mechanics centered on the uniqueness of the results of the matrix theory of quantum mechanics (see Born-Jordan-Heisenberg 1925-1926). All this led to the idea that developing quantum mechanics entails the search for new representations of the canonical commutation relations. Later the famous Gelfand-Naimark-Segal(GNS) construction [26], relating every representation of the CCR to an expectation value, or a state of the CCR and vice-versa, opened the gate for the state approach to the ground and/or equilibrium states of all physical models and, in particular, also for boson systems.

The physics of BEC was quickly realized to be related to the phenomenon of superfluidity. One understood in short order that superfluidity could only be explained

A.F. Verbeure, *Many-Body Boson Systems*, Theoretical and Mathematical Physics, DOI 10.1007/978-0-85729-109-7_1, © Springer-Verlag London Limited 2011

on the basis of an interacting Bose gas [20, 21]. In short, the spectrum of the free Bose gas did not fit with the property of superfluidity. The first step to overcome this difficulty was the introduction of an interaction, the mean field or imperfect Bose gas, which conserves nevertheless the solvability of the model. However, this model, whose spectrum is identical to that of the free Bose gas, did not produce the right solution which produces a spectrum giving an explanation for the phenomenon of superfluidity.

In this context the so-called Bogoliubov model, sometimes called the weakly interacting Bose gas [169, 22, 169], was conceived. This model takes into account more interaction terms but without losing its interesting property of exact solvability. The basic ingredients of this model in terms of states on the CCR algebra of the boson observables have been analyzed rigorously in [8, 9, 7]. Later the boson-pairing model was introduced as an exercise towards a further refinement of the Bogoliubov model. Theoretical work on this model resulted in some intriguing questions, such as whether two types of condensation occur simultaneously or not. The question addresses the co-existence of a boson-pair condensation and the standard one-particle ground state condensation. The other question deals with whether a spectral gap appears in the spectrum of the elementary excitations [58, 108, 90, 82, 142].

After a number of quiet years came the great and important year, 1995, now considered the year BEC was experimentally discovered for the ideal Bose gas. We refer of course to what is called Bose condensation for trapped alkali metals, research done by E. Cornell and C. Wieman at JILA, Boulder, R. Hulet at Rice Univ. Houston Texas and W. Kelterle at MIT, Cambridge Mass, and by many people active in the enormous scientific activity presently taking place in this field.

Moreover, many boson models have been heavily studied in the literature, not only in statistical mechanics but also in field theory. In all aspects, model studies of boson systems have always represented a large part of the activity in this research field. Therefore considerable attention is devoted in this book to the discussion of solvable models.

A characteristic feature of all solutions of solvable boson models is that they share a common property for the equilibrium states as well as for the ground states; namely, they are completely determined by their one- and two-point correlation functions. Higher-order correlation functions are expressed in terms of these one- and two-point functions. Because of their similarities with the Fock state, the ground state of the free boson gas, such states are called generalized free or quasi-free states. This class of states was intensively studied in the 1960s and 1970s. These quasi-free states are now in a dispersed order commonly used as the ideal theoretical laboratory in which we can perform tests of all kinds. Although an extensive literature about these states now exists, the intensely pure mathematical analysis of quasi-free states was apparently much too technical and general to be practical for the study of boson systems by theoretical physicists.

These notes are intended to remedy to this situation. An effort is made to present a less technical and more accessible presentation of the subset of quasi-free states within the set of all states. A proper definition of the notion of solvable model is provided. The relation with quasi-free states is also clarified. Also their basic properties and a number of their applications are explicitly discussed. On the other hand, only those quasi-free states suitable to the study of space-translation invariant or space-homogeneous boson systems are discussed. Special attention is paid to the explicit form of the variational principle of statistical mechanics for all solvable boson models. This principle is extensively discussed. We deal explicitly with the question as to why the general variational principle of statistical mechanics for solvable boson models can be reduced to a tangible, manageable variation over the set of quasi-free states. This property is expressed in a form practical enough to obtain directly and explicitly the corresponding quantum Euler-like equations of this special variational principle.

Apart from the special attention to solvable models, the universal problem and the universal properties of equilibrium and non-equilibrium systems for fully interacting boson systems is given ample space and a primordial place in this book. The emphasis is more on rigorous and conceptual results about these interacting systems and less about approximate descriptions and/or hard computational matters.

As is well known, the situation of equilibrium statistical mechanics is far more developed than the non-equilibrium case. As for the former, there are many equilibrium conditions formulated which can be found in the literature and which can be called *quasi-equivalent* or *equivalent up to small variations*. In fact there exists an infinity of such equilibrium conditions. We limit ourselves to two equilibrium conditions formulated for quantum systems, each of them directly expressing a clear physical property characterizing completely the equilibrium states. The first one is the commonly known variational principle of statistical mechanics. The second one is the energy-entropy balance criterion for equilibrium, which can be considered the quantum Euler equations version of the variational principle for quantum systems. However the range of applicability of the two criteria is not the same. The second one has a wider range of applicability than the variation principle, which is not so well suited whenever we deal with non-homogeneous systems simply because of the difficulties in taking thermodynamic limits. The choice of these two conditions is motivated by the fact that both of them have a direct and generally acceptable physical interpretation as equilibrium criteria holding in the thermodynamic limit. The variational principle expresses the property of an equilibrium state being characterized as a state minimizing the free energy density functional.

For the second criterion of equilibrium, the energy-entropy balance criterion (which is also valid for non-homogeneous systems), we can simply say that equilibrium states are characterized by the basic property that under any perturbation their energy change is always majoring their entropy increase. It is an attractive concept to realize that this simple physical idea yields indeed a full characterization of an equilibrium state. As already pointed out, those of us with some experience in the study of equilibrium conditions will also point out that, apart from our two characterizations, the literature (see e.g. [26]) contains many more characterizations of equilibrium.

Looking only at the mathematical physics literature, in particular in the field of the algebraic approach to quantum statistical mechanics, the most popular characterization of equilibrium so far is formulated by means of the so-called KMS (Kubo-

Martin-Schwinger) conditions. Because they were the first quantum criteria specifying equilibrium in the thermodynamic limit, the KMS conditions have also become the best known criteria for specifying quantum equilibrium states.

The influence of the KMS paper [73] on the mathematical physics community has been especially significant. The KMS conditions have been considered to be the quantum mechanical counterparts of the DLR (Dobrushin-Lanford-Ruelle) equations for classical systems. The DLR equations had a direct physical interpretation because of their formulation in a full probabilistic language in terms of expectations and conditional expectations.

Around that period (1970) probability theory became the theory of common sense, popular even far beyond the narrow limits of physics and mathematics. Statistics entered daily life. The truth in social and political life starts getting determined by "cleverly chosen" distributions for a number of "objectively" measured random variables. In this atmosphere the DLR approach got very easily the realm of a good intuitive understanding of classical equilibrium and therefore was better digested by a larger crowd of mathematicians and physicists. On the other hand, the quantum mechanical KMS conditions did not preach such a self-evident physical interpretation of equilibrium. Rather, they came across as a pure mathematical property, which by the way relies on numerous mathematical technical concepts simply to formulate decently all conditions.

Because of these aspects, the formulation of the KMS conditions significantly boosted many activities in pure mathematics, in particular functional analysis. The formulation of the KMS conditions requiring a number of technical assumptions is making them less useful for direct applications to physical systems. This is particularly the case for the applications to boson systems, because in these systems the basic observable quantities are unbounded operators. For all these reasons, this equilibrium criterion has not been very popular within the larger community of physicists so far, except for the mathematical physics community.

Nevertheless we are convinced that there are good reasons to popularize and promote a number of general basic ideas of this algebraic approach, which are interesting but less known. These ideas are leading to new ways of looking at standing problems, new technologies for solving old problems, and finally new applications leading to new exact results. Of course we focus mainly on exact results for boson systems. All this should be done using a language which tries to bridge mathematics and physics. We aim for a presentation of the many-body boson problem highlighting the status of some exact and relevant results obtained so far in this field half a century later, although less known by the larger physicists community.

This book attempts to unify the presentation of basic concepts, positioning clearly the situation of solvable quantum boson models versus the basic universal exact results and the complexity of the problems. More generally, this book establishes links between key concepts such as equilibrium conditions and non-equilibrium aspects, spontaneous symmetry breaking and Bose-Einstein condensation, the notion of condensate equations and their position within the basic theory, quantum fluctuations and the phenomenon of bosonization, and reversible and irreversible dynamical systems. It presents tools needed to analyze qualitatively and quantitatively different types of models. The conceptual aspects of the text may very well produce a more universal dimension to the subject matter than presently understood.

We mentioned that the notion of quasi-free states is connected with solvable boson models. We make clear that this notion, however, extends far beyond the restricted environment of solvable microscopic boson systems. We point out that boson quasi-free states emerge in a natural way in all locally interacting systems as the canonical states of the normal quantum fluctuations of local quantum observables. We also included a chapter about the notion of the quantum fluctuation operator and we discuss its properties. The quasi-free states are the quantum mechanical counterparts of the classical Gaussian distributions [63]. Furthermore we show that quantum fluctuation operators always satisfy boson canonical commutation relations. The socalled bosonization of all kinds of micro-systems fits within this chapter.

We tried to formulate the contents of this book in a language aimed at a technical level accessible to the greater community of theoretical physicists. In some places the extreme technical mathematical generality in the formulation is deliberately left open to give us the occasion of giving their own inspiration and even interpretation to the material. Whether we like it or not, mathematical technology sometimes kills enthusiasm within sound color. Therefore practicing physicists reading this book are warned that we explain mainly simple facts about bosons, only trying to make the phenomena understandable from the microscopic point of view. They should expect neither high-tech stands nor hype. Students, graduate students, and inexperienced individuals reading this book are warned as well; we keep it easy because professors may also read this book.

2

Bose systems

2.1 Generalities

In physics the concrete and traditional approach to Bose systems is to start with the Fock Hilbert space of vector states which we denote here by $\mathfrak{F}$, with a scalar product given by $(.,.)$.

Consider $L^2(\mathbb{R}^d)$, the space of square integrable functions on $\mathbb{R}^d$, $d = 1,2,...$ is the dimension of the system under consideration. Denote by $\mathscr{S}$ the space of nice (infinitely differentiable, rapid decrease) test functions in d dimension, as a subspace of $L^2(\mathbb{R}^d)$. All functions of the set $\mathscr{S}$ stand for the wave functions of the individual boson particles in the system. They are also called the *one-particle wave functions*. One considers the one-particle creation and annihilation operators with wave functions any $f,g \in \mathscr{S} \subset L^2(\mathbb{R}^d)$. The creation operator is given by $a^*(f) = \int dx f(x) a^*(x)$ acting on the space $\mathfrak{F}$. The annihilation operator is the adjoint operator of the creation operator and given by $a(f) = \int dx \overline{f}(x) a(x)$. These two operators satisfy the usual *canonical commutation relations(CCR)*

$$[a(x), a^*(y)] = \delta(x-y), \quad [a(x), a(y)] = 0 \tag{2.1}$$

for any $x, y \in \mathbb{R}^d$. This leads immediately to the mathematically more complete but somewhat less popular form of the canonical commutation relations: $\forall f, g \in \mathscr{S}$ holds

$$[a(f), a^*(g)] = (f,g), \quad [a(f), a(g)] = 0 \tag{2.2}$$

It is assumed that there exists a particular normalized vector Ω in the Fock Hilbert space $\mathfrak{F}$ such that it is annihilated by all $a(x)$ and hence that: $\forall f \in \mathscr{S}$ holds

$$a(f)\Omega = 0 \tag{2.3}$$

Because of this property the vector Ω is called the *vacuum vector* of the Fock space. On the other hand the Fock space $\mathfrak{F}$ itself is taken to be the Hilbert space linearly generated by all vectors of the following set: $a^*(f_1)a^*(f_2)...a^*(f_n)\Omega$ for all $f_i \in \mathscr{S}$ and for all natural numbers $n \in \mathbb{N}$.

A.F. Verbeure, *Many-Body Boson Systems*, Theoretical and Mathematical Physics, DOI 10.1007/978-0-85729-109-7_2, © Springer-Verlag London Limited 2011

The elements of this Fock Hilbert space are called the Fock space wave functions of the boson systems.

In this book, we use the word *state* of a boson system for each expectation valued map see Eq. (7.1) or simply for each expectation. Later we give a more detailed, more precise and more general description of this map. Nevertheless starting from the Fock Hilbert space context we describe already in a bit more details the concept of state. Let Ψ be any normalized vector, i.e. any wave function or vector of the Fock Hilbert space, then the expectation values of the type $\omega_\Psi(A) = (\Psi, A\Psi)$, where A stands for any observable of the boson system, define a *state* ω_Ψ, an expectation valued map, a map of the observables into the complex numbers.

As always the set of observables of a boson system is given by the *algebra of observables*, the algebra generated by the creation and annihilation operators.

Of course, any other representation space of the boson observables $a(x), a^*(x)$, different from the Fock space representation, yields other sets of vector states and hence other sets of states or expectation valued maps.

As said above each boson observable is expressed as a function of the boson creation and annihilation operators. In particular, each physical model is defined by giving explicitly its energy observable, called its *Hamiltonian*. For any two-body inter-particle interaction potential v, the general boson model takes the following explicit form in any finite spacial volume V, a subset of $\mathbb{R}^d$:

$$H_V = \int_V dx \frac{1}{2m} \nabla a^*(x) . \nabla a(x) + \frac{1}{2} \int_V dx dy\, a^*(x) a^*(y) v(x-y) a(x) a(y) \qquad (2.4)$$

where m is the mass of the particle. For simplicity of notation we put Planck's constant $\hbar = 1$. The Hamiltonian consists of the sum of two terms. The first term represents the kinetic energy of the individual boson particles of the system. The second term represents the interaction energy between the individual particles. The stability of the model requires that the Hamiltonian operator Eq. (2.4) acting on the Fock Hilbert space is bounded from below. This stability requirement puts some conditions on the potential v. Also this point is discussed later in more details.

Clearly, in order to define this Hamiltonian completely in each finite volume V, one has to specify correctly the derivatives appearing in the expression Eq. (2.4) at the boundaries of the volume. One speaks about specifying the *boundary conditions*. The problem of boundary conditions for quantum systems is a non-trivial affair. An interesting account about this topic is found in [56]. Because of the trickiness of the boundary conditions, physicists like the scheme of the *periodic boundary conditions*. In this case one takes for the finite volume sets V, the cubic boxes with sides of length L. By the same symbol V we denote as well the volume $V = L^d$ of these sets. For any such box V, one considers also the dual set V^* of the finite volume set V, namely as the set

$$V^* = \{k \in \mathbb{R}^d \,|\, k = \frac{2\pi}{L} n, n \in \mathbb{Z}^d\} \qquad (2.5)$$

For each $k \in V^*$, denote $\varepsilon_k = k^2/2m$ and $a_k^* = V^{-1/2} \int_V dx\, e^{ik.x} a^*(x)$. Remark that the operator a_k depends in general on the volume, a property which is not indicated in the notation. In terms of this notation is the Hamiltonian Eq. (2.4) rewritten in the following form

$$H_V = \sum_{k \in V^*} \varepsilon_k a_k^* a_k + \frac{1}{2V} \sum_{k,k',q} v(q) a_{k+q}^* a_{k'-q}^* a_{k'} a_k \tag{2.6}$$

where $v(q)$ stands now for the Fourier transform of the potential v.

A many-body physicist is interested to know basically everything about all the properties of the systems defined by the Hamiltonian Eq. (2.6). He is interested in the spectra, the dynamics, the ground states, the equilibrium states, etc ... etc. Unfortunately the nasty aspect of this general model is that it is in general not a solvable system. Not its ground states nor its equilibrium states (see next Chapter) have ever been rigorously computed for a non-trivial potential function v.

Essentially, a system is often called solvable if one can construct an expectation valued map or a state $\widetilde{\omega}$ which is the ground state or the equilibrium state at a certain temperature of the system H_V for some volume V or in the limit V (or L) tending to infinity. In the rest of the book we use the notation $\lim_V$ in order to indicate this limit. This limit is always referred to as the *thermodynamic limit*. A more precise and complete technical definition of solvability comes later.

For a boson system, as all observables are build up by means of the creation and annihilation operators and because of the canonical commutation relations, a particular expectation valued map or a state, say $\widetilde{\omega}$, is known if one knows all its *correlation functions*, i.e. if one knows all the expectation values of the type

$$\widetilde{\omega}(a^*(f_1)...a^*(f_n)a(g_1)...a(g_m)) \tag{2.7}$$

for all functions $f_i, g_j \in \mathscr{S}$ and for all pairs $n, m \in \mathbb{N}$. One should realize that in order to know the state one has to know or to specify an infinity of correlation functions, one for each pair (n, m). The expression given by (2.7) is called the *(n,m)-correlation function* or the *correlation function of order n+m*. Clearly there are in general infinitely many different correlation functions.

This infinity can make the problem of solving an interacting boson system infinitely difficult. This is the origin of the expression, a boson system with a non-trivial potential is in general non-solvable. This infinity is also the main reason for the fact that the search for solvable models is a genuine occupation for many researchers working in many-body boson systems.

In the physics literature one finds many approximation procedures which consist of reducing this infinity of different correlation functions of a ground state or of an equilibrium state to a finite number of independent ones. Many of them consist of different decoupling suggestions of arbitrary (n,m)-correlation functions Eq. (2.7). For instance, a lot of these proposed approximation procedures are of the type of assuming that all higher order correlation functions, say of order larger than $n + m$, can be expressed in terms of those of lower orders, say less than $n + m$. It must be remarked that, on the basis of the theorem of Marcinkiewicz [146], many of them are erroneous in the sense that such decoupling procedures may contradict the positivity (see further) of the state $\widetilde{\omega}$. In fact this theorem tells us something very important. It tells us that the only valid decoupling procedure is the following. If the decoupling holds for all correlation functions from some order $n + m$ on, then the decoupling holds for all correlation functions of all orders $n + m > 2$. This means that the only mathematically rigorous and physically meaningful decoupling, not contradicting the positivity of the state $\widetilde{\omega}$, is the decoupling whereby the state is given in terms of the one- and two-point functions, which are in finite number and given by the following: for all f, g: $\widetilde{\omega}(a(f))$, $\widetilde{\omega}(a^*(f)a(g))$, $\widetilde{\omega}(a(f)a(g))$. In other words this means that in general, any state is determined by a fixed finite number, maximum three, of correlation functions or by an infinity of them. Or stated in general, any state is, or determined by its one and two-point correlation functions, or by an infinity of different correlation functions. All intermediate cases are contradicting the positivity property of the state as an expectation valued map.

Of course, the most representative example of a state determined by its one and two-point functions is the so-called the *Fock state* ω_F or the Fock expectation valued map $\omega_F(A) = (\Omega, A\Omega)$, satisfying (2.3). It is an instructive student exercise to check the decoupling property for the Fock state. As is well known, the Fock state yields the ground state (see also further on) of the free boson gas (see Eq. (2.4) with $v = 0$).

Furthermore any state satisfying the correct decoupling procedure which is described above will be called a *quasi-free state* and any boson model whose ground state is given by such a quasi-free state shall be called a *solvable model*. More complete and more elaborate definitions of these two important notions are explained in more details below.

The rest of this chapter is devoted to the essentials of the boson canonical commutation relations, to its set of space homogeneous states, its gauge invariant states and its subset of quasi-free states. The latter ones will be very practical for the formulation of the variational principles of statistical mechanics for solvable models. The more mathematically minded reader can find more general and more sophisticated treatments of some of the material in [131] and [26].

2.2 CCR and boson fields

We introduced the creation and annihilation operators $a^\sharp$ acting on the Fock space $\mathfrak{F}$ where the symbol $\sharp$ refers either to the creation operator or to the annihilation operator. In order to define the total set of states as well as the set of all quasi-free states it is sometimes handy to work with the notion of boson fields.

The *boson field* is given by the map $b : f \in \mathscr{S} \subseteq L^2(\mathbb{R}^d) \to b(f)$, where

$$b(f) = a(f) + a^*(f) \tag{2.8}$$

Clearly each $b(f)$ is a self-adjoint linear operator on the Fock Hilbert space. In the physics literature, $\mathscr{S}$ is also called a space of test functions consisting of the infinitely differentiable functions with rapid decrease at infinity. Remark that the creation operators are complex linear on the space of test functions $\mathscr{S}$, the annihilation operators are complex anti-linear, and therefore the fields are only real linear in their arguments, one computes e.g. $b(if) = i(-a(f) + a^*(f))$ and therefore $a(f) = \frac{1}{2}((b(f) + ib(if))$.

The *canonical commutation relations* (2.2) translated in terms of the fields take now the form

$$[b(f), b(g)] = 2i\sigma(f, g) \tag{2.9}$$

with $\sigma(f, g) = \Im(f, g)$. Remark that σ is a real bilinear antisymmetric form on the real test-function space $\mathscr{S}$. Such a form on a real vector space is called in general a *symplectic form* and any real linear space equipped with a symplectic form is called a *symplectic space*. Hence the couple $(\mathscr{S}, \sigma)$ realizes a symplectic space. The reader understands from Eq. (2.9) that the canonical commutation relations in terms of the fields are completely determined solely by such a symplectic form σ. This conclusion holds as well for the next formulation of the commutation relations.

Clearly, working with the field operators as the generators of all observable quantities or working with creation and annihilation operators are equivalent procedures.

It is also clear that both sets of generators, as well the fields as the creation-annihilation operators, consist of unbounded operators. For many mathematical manipulations and argumentations working with bounded operators has its technical mathematical advantages. Therefore, but also for other technical reasons, Weyl proposed to use the following unitary operators as the generators of all the observable quantities of boson systems. For any $f \in \mathscr{S}$ the corresponding *Weyl operator* is given by the unitary operator

$$W(f) = exp\{ib(f)\} \tag{2.10}$$

Using the well known Baker-Campbell-Hausdorff formula argument, telling that for the operators X and Y, both commuting with the commutator $[X, Y]$, holds

$$e^X e^Y = e^{X+Y} e^{\frac{1}{2}[X,Y]}$$

one derives from (2.9) that the canonical commutation relations Eq. (2.2) in terms of the Weyl operators get the following form:

$$W(f)W(g) = W(f + g)e^{i\sigma(f,g)} \tag{2.11}$$

It should be clear that algebraically, i.e. up to a number of topological aspects which we disregard here, one has now the option to work with three different presentations of the *boson algebra of observables* $\mathfrak{A}$, namely:

(i) the algebra generated by the creation and annihilation operators and the unit operator,

(ii) the algebra generated by the fields and the unit operator, or

(iii) the algebra generated by all the Weyl operators.

The latter one is of course the most suitable one for the research in pure mathematical physics and is very much used in the field called the algebraic approach to statistical mechanics and field theory. The two other presentations are mostly used by theoretical physicists as is also well known. A rather complete account of all this can be found in [26].

So far for the *algebra of observables*, which we denote simply by $\mathfrak{A}$ for any of the tree settings.

2.3 States and Quasi-Free States

We discuss the set of states on the algebra of observables 𝔄. As indicated above, a state on the algebra 𝔄 is an expectation valued map.

Definition 2.1. *The mathematical properties of any state ω are the following: it is a normalized-to-one, linear, positive form or functional on the algebra of observables* 𝔄. *More explicitly, the state ω maps each observable $A \in$ 𝔄 *into its expectation value which is in general a complex number $\omega(A)$ with the properties*
(i) normalization: $\omega(1) = 1$
(ii) linearity: for each pair A, B of observables and each pair λ, μ of complex numbers one has $\omega(\lambda A + \mu B) = \lambda \omega(A) + \mu \omega(B)$
(iii) positivity: for each observable A holds the positivity of the expectation value $\omega(A^*A) \geq 0$

It is perhaps important to realize that these are indeed the essential properties that a state ω should have in order to give to the expression $\omega(A)$ the interpretation of the expectation value of an observable A as we learned about in our undergraduate lectures. For these reasons it is clear that the notion of state is indeed a mathematical formalization of the notion of expectation valued map. The notion of state has a direct link to the notion of observation values in quantum physics and is therefore much more to the point than the notion of wave function. Remark that in our language the notion of state is not the same as the one which one finds in standard books on quantum mechanics, where a state is a vector, called the wave-function, an element of a Hilbert space, which itself can be referred to as the set of vector-states. The link between the notion of vector-state or wave function, and our more general notion of state, is mathematically realized by the so-called *GNS-construction* (see [26] and (7.1)), which is a very general and important theorem telling us the following. Let ω be any state on the algebra of observables 𝔄, then there exists a representation π of the canonical commutation relations algebra acting on a Hilbert space $\mathcal{H}$ and a special vector, called *cyclic vector*, Ω in $\mathcal{H}$ such that $\omega(X) = (\Omega, \pi(X)\Omega)$ for any observable X. We should immediately remark that the Hilbert space $\mathcal{H}$ needs not to coincide with the original Fock space. One knows more. One knows that in most physically interesting cases they do not coincide. The reader does understand that our notion of state does create a generality concerning expectation values which is going far beyond the Fock space. It turns out that this generality is necessary in order to understand the most interesting phenomena about systems with a large(=infinite) number of degrees of freedom. This is the main reason for our choice of working with the notion of state or expectation valued map in stead of with wave functions.

Denote by $\mathscr{E}$ the *set of states* on the boson observable algebra. First of all it is interesting to remark that this set $\mathscr{E}$ is a convex set, i.e. for each pair of states ω_1, ω_2 and each real number λ in the interval $[0, 1]$, also the convex combination $\omega = \lambda \omega_1 + (1 - \lambda) \omega_2$ is again a state. It is clear that the value of λ has the physical interpretation of the concentration of the state ω_1 in the state ω and $1 - \lambda$ the concentration of ω_2 in the state ω. This remark leads to the following notions. A state is called a *pure state* or an *extremal state*, if it is not possible to write the state as a non-trivial convex combination of two other different states. Hence the state ω is pure if $\omega = \lambda \omega_1 + (1 - \lambda) \omega_2$ implies $\lambda = 0$ or $\lambda = 1$, and/or implies $\omega_1 = \omega_2$. A state is a *mixed state* if it is not a pure state. Of course the notion of convex combination of finite sums extends straightforwardly to infinite sums and even to integrals of states. In particular the following situation will be relevant for us. Consider any convex set S of a vector space equipped with a probability measure μ (i.e. for $\lambda \in S, \mu(\lambda) \geq 0, \int d\mu(\lambda) = 1$) defined on S. Let $\{\omega_\lambda\}$ be a set of states labeled by the parameter λ, then also the integral $\omega = \int d\mu(\lambda)\omega_\lambda$ is again a state, because the set of states $\mathscr{E}$ is a convex set.

Now we look for a general but practical definition of the set of states for boson systems also with the intention to formulate clearly the particular subset of states which will be called the set of quasi-free states. For these aims the Weyl formulation is very suitable.

Let ω be an arbitrary state on the Weyl algebra $\mathfrak{A}$. This state is known or well defined, if for all $f \in \mathscr{S}$, all the expectation values $\omega(W(f))$ are known, or if the expectation values of all Weyl operators are known. A straightforward classical computation (see e.g. [146]) yields the expression of the expectation values in terms of the field correlation functions

$$\omega(W(f)) = \omega(e^{i\lambda b(f)}) = \sum_{n=0}^{\infty} \frac{i^n \lambda^n}{n!} \omega(b(f)^n) \qquad (2.12)$$

$$= \exp\{\sum_{n=1}^{\infty} \frac{i^n \lambda^n}{n!} \omega(b(f)^n)_t\}$$

where the so-called *truncated correlation functions* $\omega(...)_t$ are defined recursively through the formula

$$\omega(b(f_1)...b(f_n)) = \sum \omega(b(f_k)...)_t...\omega(...b(f_l))_t \qquad (2.13)$$

where the sum is over all possible partitions $(k,...),(...),...(...l)$ of the set $\{1,...,n\}$, with the order within each of the clusters carried over from the left to the right hand side. Because of this definition, one calls each $\omega(b(f_1)...b(f_n))_t$ the truncated correlation function of order n. In the physics literature the word "connected" is also sometimes used in stead of the connotation "truncated".

The formula Eq. (2.12) expresses that the state is completely defined if one knows all its truncated correlation functions and vice versa.

In this Weyl formulation the basic properties of a state are now explicitly given by: let $A = \sum_i c_i W(f_i)$ be any arbitrary element of the algebra of observables, then
(i) normalization: $\omega(W(0)) = \omega(1) = 1$
(ii) linearity: $\omega(A) = \sum_i c_i \omega(W(f_i))$
(iii) positivity: $\omega(A^*A) \geq 0$

This completes the definition of a general state on the algebra of boson observables in terms of the field correlation functions or equivalently in terms of the truncated field correlation functions.

Now we are able to identify a very special class of states, namely the set of quasi-free states.

Definition 2.2. *A state ω of the boson algebra of observables is called a quasi-free state, also written qf-state, if all its truncated correlation functions of all orders $n > 2$ vanish. This has the immediate consequence that all its $(n > 2)$-correlation functions are expressed in terms of those of orders $n \leq 2$.*

From the formula Eq. (2.12) it follows that the most general quasi-free state is completely determined by its one- and two-point correlation functions and therefore gets the following simpler explicit form:

$$\omega(W(f)) = \exp\{i\omega(b(f)) - \frac{1}{2}\omega(b(f)b(f))_t\} \tag{2.14}$$

We denote by $\mathfrak{Q}$ the set of all quasi-free(qf) states. Some authors call the set of quasi-free states also the set of generalized free states.

Remark that, if one takes any real linear functional $\chi : f \to \chi(f)$ on the test function space $\mathscr{S}$, then any such functional defines a *canonical transformation* Eq. (7.3) τ_χ, i.e. a one-to-one transformation mapping the observables onto the observables leaving the canonical commutation relations (CCR) invariant. It is acting on the boson algebra $\mathfrak{A}$ in the Weyl form as follows

$$\tau_\chi(W(f)) = e^{i\chi(f)}W(f) \tag{2.15}$$

together with the rules that τ_χ is linear, conserves all products as well as the *-operation. Note that the composition of any state ω with the canonical transformation τ_χ, i.e. that $\omega \circ \tau_\chi$, is again a state.

It is immediately checked from Eq. (2.15) that the action of this transformation is nothing else but translating the boson fields with a scalar quantity, namely: $\tau_\chi b(f) = b(f) + \chi(f)$. This follows directly from the formal computation

$$\frac{d}{d\lambda}\tau_\chi(W(\lambda f))|_{\lambda=0} = e^{i\lambda\chi(f)}W(\lambda f)|_{\lambda=0}$$

Take any qf-state ω Eq. (2.14), then the composition $\widetilde{\omega} = \omega \circ \tau_\chi$ is not only again a state, one readily checks that it is again a qf-state. In particular if one chooses $\chi(f) = -\omega(b(f))$ then the one-point function of the new state $\widetilde{\omega}$ vanishes. Moreover the two-point truncated function is left invariant for the transformation τ_χ, i.e. $\widetilde{\omega}(b(f)b(g))_t = \omega(b(f)b(g))_t$. Therefore, up to such a canonical transformation, one can continue the analysis of the set of qf-states with the set of qf-states restricting ourself to those with vanishing one-point function. In this case the definition of qf-state Eq. (2.14) reads as follows

$$\omega(W(f)) = \exp\{-\frac{1}{2}\omega(b(f)b(f))\} \tag{2.16}$$

because in this case $\omega(b(f)b(g))_t = \omega(b(f)b(g))$.

Denote $s(f,f) = \omega(b(f)^2)$ then Eq. (2.16) becomes

$$\omega(W(f)) = \exp\{-\frac{1}{2}s(f,f)\} \tag{2.17}$$

Denote also by $s(f,g)$ defined on $\mathscr{S} \times \mathscr{S}$, the real bilinear symmetric extension of $s(f,f)$ on $\mathscr{S}$. By differentiating twice $\omega(W(\lambda f)W(\mu g))$ with respect to λ and μ at zero, one obtains the two-point function for the fields in the following form

$$\omega(b(f)b(g)) = s(f,g) + i\sigma(f,g) \qquad (2.18)$$

The positivity condition applied to the qf-state ω becomes: for all $A = \sum c_i W(f_i)$ holds

$$0 \leq \omega(AA^*) = \sum_{j,k} c_j \overline{c_k}\, \omega(W(f_j - f_k)) \exp(-i\sigma(f_j, f_k))$$
$$= \sum_{j,k} (c_j e^{-\frac{1}{2}s(f_j,f_j)})(\overline{c_k} e^{-\frac{1}{2}s(f_k,f_k)}) \exp(s(f_j,f_k) - i\sigma(f_j,f_k))$$
$$= \sum_{j,k} d_j \overline{d_k}\, \exp(s(f_j,f_k) - i\sigma(f_j,f_k))$$

where the parameters d_k are immediately identified from the second line. Use the following general matrix property which is straightforwardly checked. If the matrices $A = (a_{i,j})$ and $B = (b_{i,j})$ are positive definite $n \times n$-matrices, then the matrix $C = (a_{i,j}b_{i,j})$ is also a positive definite matrix (see e.g. [131]). This property yields immediately the proof of the fact that the positivity of the qf-state is equivalent to the positive definiteness of the two-point function Eq. (2.18). Expressed in words, the positivity of a qf-state ω restricted to the monomials in the fields of order two, is necessary and sufficient for the full positivity of the qf-state.

All field correlation functions can as well be expressed in terms of the creation and annihilation operators (n,m)-correlation functions Eq. (2.7) and vice versa. Therefore one can express as well, and equivalently, this positivity in terms of the creation and annihilation operators $a^\sharp$ which are complex linear, respectively anti-linear in the test functions. The positivity of the quasi-free state ω is therefore expressed by: $\forall f,g \in \mathscr{S}$, considered now as a complex linear space, the positivity of the state becomes

$$\omega^*((a(f)+a^*(g))(a(f)+a^*(g))^*) \geq 0 \qquad (2.19)$$

This is indeed the necessary and sufficient condition for the positivity of the state.

The next step in the analysis of the states of boson systems is by introducing the parameterizations of the truncated two-point functions of any state ω by means of operators. In the following we consider states for which the truncated two-point functions are determined by the (unbounded) operators R and S acting on the space $\mathscr{S}$, and which are defined as follows

$$\omega(a(f)a^*(g))_t = (f, Rg); \quad \omega(a(f)a(g))_t = (f, S\overline{g}) \qquad (2.20)$$

where the symbol $\overline{g}$ stands for the complex conjugate of g. Clearly the symbol $(.,.)$ stands for the scalar product on $L^2(\mathbb{R}^d)$.

It is important to note that this operator presentation holds for the truncated two-point correlation functions of any state, being quasi-free or not. Moreover, an identical operator representation of any truncated (n,m)-correlation function (see

Eq. (2.7) or Eq. (2.13)) can be obtained by an operator mapping any dense subspace of $L^2(\mathbb{R}^{dn})$ into $L^2(\mathbb{R}^{dm})$.

In particular for quasi-free states, as all higher $(n+m > 2)$ order truncated correlation functions vanish, one can rewrite the full positivity condition Eq. (2.19) of a quasi-free state solely in terms of the operators R and S, defined in Eq. (2.20). The reader realizes that the positivity conditions of a general state involve however all correlation functions of all orders.

Before continuing the analysis of the positivity conditions, it may be instructive to illustrate first the material by making a small intermezzo presenting a couple of well known examples of states and quasi-free states in terms of their operator presentation. After that we concentrate ourself onto the subclasses of all states which are space homogeneous or space translation invariant and gauge invariant.

Examples of boson systems states

The notion of quasi-free state is not so mysterious as it may sound. The examples used all around in the physics literature are daily matters. As already mentioned, the best known example of a quasi-free state is the *Fock state* ω_F given by: for all $f \in \mathscr{S}$,

$$\omega_F(W(f)) = (\Omega, W(f)\Omega) = e^{-\frac{1}{2}(f,f)} \tag{2.21}$$

where Ω is the vacuum wave vector of the Fock Hilbert space Eq. (2.3). The GNS-representation space (7.1) of the Fock state is the Fock Hilbert space $\mathfrak{F}$ and the cyclic vector is the vacuum vector Ω. In other words the Fock state is the expectation valued map determined by the Fock vacuum vector. From Eq. (2.21) it follows that the defining operators (R,S) Eq. (2.20) of the Fock state ω_F are given by the operators $R = 1$ and $S = 0$. Furthermore the one-point function of the Fock state vanishes.

A subset of the set $\mathfrak{Q}$ of qf-states, is the set of so-called *coherent states* associated to the Fock state ω_F. Denote this set of coherent states by $C(\omega_F)$. This set of states is given by all states $C(\omega_F) = \{\omega_h ; h \in \mathscr{S}\}$ where the ω_h are defined by the formulae

$$\omega_h(W(f)) = (W(h)\Omega, W(f)W(h)\Omega) = (\Omega, W(f)\Omega)e^{i2\sigma(f,h)} = \omega_F \circ \tau_\chi(W(f)) \tag{2.22}$$

with τ_χ again the canonical transformation Eq. (7.3) of the field translations with $\chi(f) = 2\sigma(f,h) = 2\Im(f,h)$. In particular it is clear that all states of the set $C(\omega_F)$ are build on the ground state wave function or the Fock vacuum wave vector Ω. It is also clear that all coherent states ω_h, associated with the Fock state, are quasi-free states of the boson systems.

However it is also clear that the notion of coherent state can be associated to any other state ω of the boson algebra of observables $\mathfrak{A}$. The set of coherent states $C(\omega)$ associated with the state ω is analogously given by $C(\omega) = \{\omega \circ \tau_\chi \mid \forall \chi(f) = 2\Im(f,h), h \in \mathscr{S}\}$. If ω is a qf-state then the set $C(\omega)$ is a subset of the qf-states. However, if ω is not a qf-state, then none of the states of $C(\omega)$ are qf-states. Hence the notion of coherence for a state is not necessary linked to the notion of quasi-freeness. If one considers the GNS-representation of the state ω: $\omega(X) = (\Psi, X\Psi)$

with cyclic vector Ψ, then this vector is the ground wave function of all states in the set $C(\omega)$. A priori this vector need not to be an element of the Fock Hilbert space $\mathfrak{F}$.

Needless to mention that the notion of coherent state has been used in many applications in physics. For instance, most of the exact results concerning the settings and the derivations of different forms of the Hartree-Fock equations [159] are obtained using the coherent state technology. It is clear that in all these applications, a main starting point consists of making the best choice for the generating state ω or of the right choice for the ground state wave function. Next to the coherent state idea and technology, there is also the wavelet state technology, which can be considered as a generalization of the coherent state technology. We do not enter here into the details about the wavelet states, because their applications are so far not too much present in many body boson physics.

Homogeneous states

The space translations are again realized by a group of canonical transformations Eq. (7.3) $\{\tau_x | x \in \mathbb{R}^d\}$ acting on the algebra of observables $\mathfrak{A}$ and are given by the maps $\tau_x(a(f)) = a(T_x f)$ where T_x is the action of translation over the distance x (7.3), $(T_x f)(y) = f(y - x)$, acting on the test function space $\mathscr{S}$.

Definition 2.3. *The set of homogeneous states or space translation invariant states is given by all states ω which satisfy the following invariance property: for all $x \in \mathbb{R}^d$ holds $\omega \circ \tau_x = \omega$.*

Let us illustrate an immediate implication of the homogeneity of a state on the correlation functions.

Using the fact that $\tau_x a_k = \frac{1}{\sqrt{V_x}} \int_{V_x} dy\, a(y) e^{-ik(y-x)} = e^{ikx} \frac{1}{\sqrt{V}} \int_{V_x} dy\, a(y) e^{-iky}$, with $V_x = V + x$, one gets $\lim_V \omega(\tau_x(a_k)) = e^{ikx} \lim_V \omega(a_k)$. Hence, if the state ω is homogeneous, one gets that for all $k \neq 0$, $\lim_V \omega(a_k) = 0$ in the thermodynamic $\lim_V$. Check that for homogeneous states ω holds in general $\lim_V \omega(a_{k_1}^* \cdots a_{k_n}^* a_{k_{n+1}} \cdots a_{k_{n+m}}) = 0$ if $k_1 + \ldots + k_n - k_{n+1} - \ldots - k_{n+m} \neq 0$.

Furthermore supposing that ω is a space translation invariant state, and using the operator representation of the two-point truncated functions Eq. (2.20), one gets that the invariance property is transported to the operators R, S with the property that both operators R and S commute with all the operators T_x, because e.g. for all x holds

$$(f, Rg) = \omega(a(f)a^*(g)) = \omega(\tau_x(a(f)a^*(g)))$$
$$= \omega(a(T_{-x}f)a^*(T_x g)) = (f, T_{-x} R T_x g)$$

and therefore $[R, T_x] = [S, T_x] = 0$. Operators with these properties are sometimes called translation invariant operators.

It is a fairly well known property [78] that if A is any such translation invariant operator, then there exists a tempered distribution on the test function space $\mathscr{S}$ with Fourier transform a function ξ such that for all functions f with Fourier transform $\hat{f}$ holds $\widehat{Af}(p) = \xi(p)\hat{f}(p)$. This means that the operator A is a simple multiplication operator. This property is a consequence of the kernel theorem for operator valued distributions and the convolution theorem for Fourier transforms. In the following, for notational convenience, we omit the notation for Fourier transforms and write simply

$$Af(p) = \xi(p)f(p) \tag{2.23}$$

In any case for homogeneous states, the two-point correlation function operators R and S are simply multiplication operators with functions which we denote by $r(p)$, respectively $s(p)$. It is easily checked from Eq. (2.20) that $r(p) = \omega(a(p)a^*(p))$ and $s(p) = \omega(a(p)a(-p)) = s(-p)$ where the $a(p)^{\sharp}$ are the usual operator valued distributions, the Fourier transforms of the creation and annihilation operators $a^{\sharp}(x)$ introduced before. For our purposes, as we consider only time reversal invariant systems, we can as well assume from now on that also the r-function is a symmetric function of its argument p: $r(-p) = r(p)$.

As all multiplication operators are two by two commuting with each other, also the operators R and S commute with each other: $[R, S] = 0$.

Analogously as for the two-point truncated correlation functions of a translation invariant state, all its truncated higher order (n, m)-correlation functions can be described by analogous multiplication operators with functions in $n + m - 1$ variables. The positivity of the state does imply of course a number of necessary and sufficient conditions on these functions. We do not write out in full details all these conditions for all the (n,m)-correlation functions of orders $n + m > 2$.

On the other hand we apply this result to the set of homogeneous quasi-free states. For these states, from the analysis given above, one can conclude that any qf-state ω with vanishing one-point function is completely and equivalently labeled by the operators R, S as well as by its associated functions r, s. Therefore the qf-state is uniquely denoted as $\omega_{(R,S)}$ as well as by $\omega_{(r,s)}$.

Now we are in a position to express the necessary and sufficient positivity conditions of a space homogeneous qf-state ω explicitly in terms of its determining operators or its corresponding multiplication functions. Writing out the positivity condition Eq. (2.19) yields: $\forall f, g,$

$$0 \le (f, Rf) + (f, S\Lambda g) + \overline{(f, S\Lambda g)} + (g, (R - 1)g)$$

where Λ is the conjugation operator with the property: $(\Lambda f, \Lambda g) = (g, f)$. Remark first that the special case $f = 0$ yields already the following condition on the operator R, namely: $R \ge 1$. In particular it follows that the operator R is self-adjoint. Using this property one gets $(\Lambda g, (R - 1)\Lambda g) = (g, (R - 1)g)$ and $\overline{(f, S\Lambda g)} = (\Lambda g, S^* f)$, and one obtains, with Λg replaced by g, the positivity conditions in the following form

$$0 \le (f, Rf) + (f, Sg) + (g, S^* f) + (g, (R - 1)g)$$

which are immediately translated into the equivalent operator or function conditions:

$$0 \le R(R - 1) - S^* S \tag{2.24}$$
$$0 \le r(p)(r(p) - 1) - |s(p)|^2 \tag{2.25}$$

This form of the positivity conditions suggests the introduction of the following non-negative function $t(p)$, defined by $t(p)^2 = r(p)(r(p) - 1) - |s(p)|^2$, expressing the full positivity of the qf-state determined by the functions (r, s). Therefore the qf-state can now better be labeled by the functions $r \geq 1, t \geq 0$ and the real number $\alpha = \arg s$. Hence in the rest of this text we may label equivalently the qf-state ω with vanishing one-point function by $\omega_{R,S}$ or as well by $\omega_{(r,t,\alpha)}$. One should remember that in this notation it is pre-supposed that the one-point function is put equal to zero. If this is not equal to zero, one should get a full parametrization only if also the one-point parameter is added to the notation. For translation invariant states we use the one-point parameter c, defined by $c\hat{f}(p=0) = \omega(a^*(f))$, which makes sense again as an immediate consequence of the homogeneity of the state. Hence a full parametrization of a qf-state looks as follows: $\omega_{c,r,t,\alpha}$.

Ergodic states

Definition 2.4. *Consider any general homogeneous or space translation invariant state ω, i.e. a state satisfying, for all $x \in \mathbb{R}^d$, the equality $\omega \circ \tau_x = \omega$. The state is called an extremal space invariant or ergodic state, if for each pair (A, B) of local observables, i.e. build up by creation and annihilation operators $a^*(f), a(f)$ with test-functions $f \in \mathscr{S}$ of finite local support, holds*

$$\lim_{|x| \to \infty} \omega(A \tau_x B) = \omega(A) \, \omega(B) \tag{2.26}$$

Notice that an ergodic state is always a space invariant or homogeneous state. The property of ergodicity of a state means that the expectation value of the product of two observables equals the product of the expectation values of each of them if one of the observables is moved far away. It means also that the state has a kind of asymptotic product property or an asymptotic independence property. Remark that ergodic states have an interesting property concerning the expectation values of space averages of observables. Indeed, let A,B,C be arbitrary local observables, and consider the expression $\lim_V \omega(AB_VC)$, where $B_V = \frac{\int_V dx \, \tau_x B}{V}$ with τ_x again the space translation over the distance x. Hence B_V is the operator B averaged over the space volume V. For any fixed finite volume V_0 holds

$$\lim_V \omega(AB_VC) = \lim_V \omega(A(\frac{\int_{V_0} dx \, \tau_x B}{V} + \frac{\int_{V-V_0} dx \, \tau_x B}{V})C)$$

Because $\lim_V (V_0/V) = 0$, the first term vanishes. Looking at the second term, take the volume V_0 such that it contains the support of the local operator C. Then the operator C commutes with the integral on the basis of the locality of the canonical commutation relations. Finally using the ergodicity of the state one gets the formula

$$\lim_V \omega(AB_VC) = \omega(AC)\omega(B) \tag{2.27}$$

Without going into too much mathematical details, this means that the space average operator $\lim_V B_V$ exists and is equal to the expectation value $\omega(B)$ of the operator B

multiplied by the unit operator i.e. $\lim_V B_V = \omega(B)\mathbb{1}$. Notice the explicit dependence of this limit on the state. In mathematics the type of limit considered in Eq. (2.27) goes under the name of weak operator limit. For more technical details about the mathematics of ergodic states one may consult [26].

Furthermore, for later applications, it is essential to mention the following relation between homogeneous states and ergodic states. In particular there is a theorem (for all mathematical details see [26] Volume I, Chapter 4) telling essentially that each space translation invariant state can be written as a convex sum of ergodic invariant states. In more explicit formulae, let $\{\omega_\lambda | \lambda \in E\}$, with λ some parameter of a convex set E, be the set of ergodic states and ω an arbitrary homogeneous state, then there exists always a probability measure μ, defined on the parameter set E, such that $\omega = \int d\mu(\lambda)\,\omega_\lambda$. This means that each homogeneous state can written as a convex combination of ergodic states. In more down to earth words, it means that if one knows all ergodic states satisfying some physical property linear on the set of the homogeneous states, one can check what it means for all the homogeneous states.

Clearly the notion of ergodicity which is introduced, is related to the non-compact invariance group, namely the complete translation group $\mathbb{R}^d$. It is clear that in the definition of ergodicity, the group $\mathbb{R}^d$ can be replaced by any infinite non-compact subgroup. Important subgroups are for instance the subgroups $\mathbb{G}$ of the translations over sublattices of the full translation group $\mathbb{R}^d$.

There exist general theorems about the question, when can a state, ergodic for the full group, be written as a convex combination of states which are ergodic for a subgroup $\mathbb{G}$. In all this it is however important to realize the fact that the notion of ergodicity is always linked to a specific infinite translation group.

All these mathematical theorems as such will not be used in the later applications. On the other hand, some of these properties will directly be derived in different physical boson systems applications.

Finally we mention that the notion of ergodic state is a rigorous mathematical notion for what in physics is sometimes called a *pure phase state*. For this reason, ergodic states are also sometimes called *extremal invariant states*. Let us mention here at least one of the applications of the decomposition theorem of an invariant state into its ergodic or extremal components, namely the decomposition into the ergodic states with respect to a non-trivial subgroup. This type of decomposition shall lead us to the main concept of the analysis of the phenomenon of spontaneously broken symmetries, discussed for boson systems in full details in Chapter Eq. (4). The reader should be aware that the notion of spontaneous symmetry breaking is an important item not only within the domain of Bose-Einstein condensation, but in fact in many more other modern theories in physics running from solid state physics over low energy nuclear physics up to high energy physics. In any case it comes over as a a phenomenon which is typical for all systems with a large or an infinite number of degrees of freedom.

Homogeneous quasi-free states and their ergodicity

Now we look for the ergodicity properties of space homogeneous qf-states. Take any homogeneous qf-state ω. In terms of the Weyl operators, we check the ergodicity condition and therefore consider the limit $|x| \to \infty$, for all local functions (functions of compact support) f, g with $g_x(y) = g(y - x)$, of the two-point functions with $A = W(f)$ and $B = W(g)$. First compute the relation for a qf-state

$$\omega(A \tau_x B) = \omega(W(f)\tau_x W(g)) = \omega(W(f))\omega(W(g)) \exp\{i\sigma(f, g_x) - s(f, g_x)\}$$

On the basis of the Riemann-Lebesgue Lemma one gets $\lim_{|x| \to \infty} \sigma(f, g_x) = 0$, as well as $\lim_{|x| \to \infty} s(f, g_x) = 0$, and the following properties of the two-point functions

$$\lim_{|x|} \omega(a(f)a^*(g_x))_t = \lim_{|x|} \int dp \, r(p)\overline{f(p)}e^{ip.x}g(p) = 0$$

$$\lim_{|x|} \omega(a(f)a(g_x))_t = \lim_{|x|} \int dp \, s(p)\overline{f(p)}e^{-ip.x}\overline{g(p)} = 0$$

Hence the two-point functions share the ergodicity property. Looking at the one-point function, one gets from the space translation invariance $\omega(\tau_x a(f)) = \omega(a(f))$ for all x, implying that the one-point function is, as pointed out above, of the form $\omega(a^*(f)) = cf(0)$ where c is some complex constant.

In any case, all this shows that each space homogeneous or space translation invariant qf-state has always the property of being an ergodic state.

In this context, it is also important to remark that any non-trivial convex combination of two different homogenous qf-states is never a qf-state. In particular this means that by taking convex combinations of qf-states one generates a new class of states which are not anymore quasi-free states. It must be recognized that a deeper characterization and understanding of the structure of this set of generated states has so far not been cleared up. It remains a challenging open problem to get a better knowledge concerning the most intrinsic mathematical and physical properties common for all convex compositions of qf-states.

Gauge invariance

Finally we consider one more group of canonical transformations Eq. (7.3) of the CCR-algebra of observables, namely the *gauge transformations group* $\{\tau_\lambda | \lambda \in [0, 2\pi] \subset \mathbb{R}\}$ Eq. (7.3), defined by the operations

$$\tau_\lambda(a^*(f)) = e^{i\lambda}a^*(f), \quad \tau_\lambda(a(f)) = e^{-i\lambda}a(f) \tag{2.28}$$

The reader checks easily that also these transformations leave the canonical commutation relations invariant. One verifies that this group is isomorphic to the additive group modulo 2π of the real numbers $[0, 2\pi]$ and therefore coincides with the compact unitary group commonly denoted by $U(1)$.

Definition 2.5. *Any boson state* ω *is called gauge invariant, if for all* $\lambda \in [0, 2\pi]$ *holds that* $\omega \circ \tau_\lambda = \omega$.

It is immediately clear from the definition that for any gauge invariant state ω all (n, m)-point correlation functions with $n \neq m$ vanish. In particular all odd-point($n + m = odd$) correlation functions vanish. In particular the one-point function vanishes.

Let us now look closer at the action of the gauge transformations on qf-states and characterize the gauge invariant qf-states. In terms of general states we confine our attention to the one and two-point functions of a state. If the state is gauge invariant, we remarked already that the one-point function vanishes. In that case looking at the two-point functions, clearly a qf-state $\omega_{(R,S)}$ transforms under a gauge transformation as follows: $\omega_{(R,S)} \circ \tau_\lambda(a(f)a^*(g)) = \omega_{(R,S)}(a(f)a^*(g))$ and $(\omega_{(R,S)} \circ \tau_\lambda)(a(f)a(g)) = e^{-i2\lambda}(\omega_{(R,S)})(a(f)a(g))$ or equivalently the operator pair (R, S) is changed into the pair $(R, e^{-i2\lambda}S)$.

Therefore the qf-state $\omega_{(R,S)}$ or in general the two-point functions of an arbitrary state are gauge invariant if and only if the operator S vanishes, i.e. if and only if the two-point functions of the state are of the form $\omega_{(R,S=0)}$, supplemented with the one-point correlation function condition $c\widehat{f}(p = 0) = \omega(a^*(f)) = 0$ or $c = 0$.

Next we derive an other general and useful property holding for arbitrary homogeneous states which can be quasi-free or not quasi-free. We show that for any translation invariant state ω with two-point functions determined by the operators (R, S), there exists always a canonical transformation τ mapping the state into a new state which has gauge invariant two-point functions, and is therefore determined by two operators of the type $(\widetilde{R}, 0)$. We determine explicitly the operator $\widetilde{R}$ as a function of the originally given state operators R and S.

This result is a generalization of a more restricted result stated in [115], where such a map is proved to exist within the set of generalized pure qf-states. Not only the existence but also the explicit construction and form of this canonical transformation τ for any initially given state ω is derived.

Lemma 2.6. *Let* ω *be any space invariant or homogeneous state with two-point truncated functions functions* $r \geq 1$ *and* $t \geq 0$. *Then there exists a canonical transformation* τ *mapping the given state into a new homogeneous state* $\omega \circ \tau$ *with two-point truncated correlation functions given by the pair of operators* $(\widetilde{R}, \widetilde{S} = 0)$. *If* $\widetilde{r}(p)$ *is the multiplication function of the operator* $\widetilde{R}$ *then it is given as a function of the original pair of functions* (r, t) *by the formula*

$$\widetilde{r} = \frac{1}{2} + (t^2 + \frac{1}{4})^{\frac{1}{2}} \qquad (2.29)$$

Applying this result to quasi-free states, all this means the following. Let $\omega_{(R,S)}$ *be a space homogeneous qf-state, there exists a canonical transformation* τ *mapping the state into a gauge invariant one, i.e.* $\omega_{(R,S)} \circ \tau = \omega_{(\widetilde{R},0)}$, *where the multiplication operator* $\widetilde{r}$ *is given by the formula above.*

Proof. If $s(p) = 0$ for p in some domain then $S = 0$ and nothing has to be proved in that domain. Therefore assume that $s(p) \neq 0$. First we apply a canonical gauge transformation such that $s(p) = |s(p)|$, i.e. we take the parameter λ in (2.28) equal to

$-\frac{1}{2}\arg s(p)$. Then one considers a second canonical transformation γ, in the physics literature called *Bogoliubov transformation* Eq. (7.3), mapping the creation and annihilation operators $a(p)^{\#}$ into new ones $\tilde{a}(p)^{\#}$, given by

$$\widetilde{a(p)} = \gamma(a(p)) = u(p)a(p) - v(p)a^*(-p) \qquad (2.30)$$

where u and v are real functions on $\mathbb{R}^d$ satisfying $u(-p) = u(p)$, $v(-p) = v(p)$ and $u(p)^2 - v(p)^2 = 1$. One checks that the new ones satisfy again the CCR-relations. Consider the two equations

$$\widetilde{r(p)} = \omega_{(\tilde{R},0)}(a(p)a^*(p)) = \omega_{(R,S)}(\gamma(a(p)a^*(p)))$$
$$0 = \widetilde{s(p)} = \omega_{(\tilde{R},0)}(a(p)a(-p)) = \omega_{(R,S)}(\gamma(a(p)a(-p)))$$

in order to express $\widetilde{r(p)}$ as a function of $r(p)$ and $s(p)$ or preferably $t(p)$. One computes explicitly the following equations from the former ones, using the symmetry of $r(p)$ and $s(p)$.

$$\widetilde{r(p)} = u(p)^2 r(p) + v(p)^2 (r(p) - 1) - 2u(p)v(p)s(p)$$
$$0 = u(p)^2 s(p) - u(p)v(p)(2r(p) - 1) + v(p)^2 s(p)$$

Looking at the second equation, one gets a quadratic equation for the function variable $x \equiv u/v$, which always takes values larger than 1 and therefore leading to a unique solution given by

$$x = \frac{r - \frac{1}{2} + \sqrt{(r - \frac{1}{2})^2 - s^2}}{s}$$

Using the relation between the functions s and t, namely $t^2 = r(r-1) - s^2$, one gets the function x expressed in the variables r, t:

$$x = \frac{r - \frac{1}{2} + \sqrt{t^2 + \frac{1}{4}}}{\sqrt{r(r-1) - t^2}}$$

Substitute this solution for x in the expressions for u and v:

$$u = \frac{x}{\sqrt{x^2 - 1}}, v = \frac{1}{\sqrt{x^2 - 1}}$$

Finally substitute this result in the first equation in order to obtain $\tilde{r}$ as a function of r and t as expressed in the Lemma.

The canonical transformation τ of the Lemma is of course given by the composition of the gauge transformation used above with the Bogoliubov transformation. This proves the first part of the Lemma. Concerning the application to the quasi-free state case, the canonical transformation τ has to be composed with the appropriate field translation canonical transformation in order to get in due case a vanishing one-point function for the new state.

Some generalities related to the physics of ergodic boson states

After the above analysis about the structure of the set of general boson states as well as about its subset of quasi-free states, we recollect here the essentials about the correlation functions for an arbitrary ergodic state as defined above, see Eq. (2.12). We add some general and important physical interpretations. All this is with an eye kept on future applications.

First we consider the one and two-point correlation functions. The general two-point truncated correlation functions are given by the formulae

$$\omega(a(f)a^*(g))_t = \omega(a(f)a^*(g)) - \omega(a(f))\omega(a^*(g)) = \int dk \, \overline{f(k)}g(k)(r(k)-1)$$

$$\omega(a(f)a(g))_t = \omega(a(f)a(g)) - \omega(a(f))\omega(a(g)) = \int dk \, \overline{f(k)g(k)}s(k)$$

and the most general one-point function by

$$\omega(a^*(f)) = \hat{f}(0)\bar{c}, \quad \omega(a(f)) = \overline{\hat{f}}(0)c.$$

The translation invariance implies indeed

$$\omega(a^*(f)) = \int_\Lambda dx f(x)\omega(a^*(x)) = \int_V dx f(x)\omega(a^*(x=0)). \qquad (2.31)$$

Hence, $\bar{c} = \omega(a^*(x=0))$. Moreover the constant c gets the following interpretation in the thermodynamic limit.

$$\bar{c} = \lim_V \omega\left(\frac{1}{V}\int_\Lambda dx a^*(x)\right)$$

and the ergodicity property of the state yields the equality

$$\rho_0 := \lim_V \omega\left(\frac{1}{V}a_0^*a_0\right) = \lim_V \omega\left(\frac{1}{V}\int_V dx a^*(x)\int_V dx a(x)\right) = |c|^2 \qquad (2.32)$$

It is important to realize that this equality does not hold if the state is not ergodic.

Formula Eq. (2.32) can also be written in the form

$$\rho_0 - \lim_V |\omega(\frac{a_0}{\sqrt{V}})|^2 = 0$$

or more explicitly as follows

$$\lim_V \frac{1}{V^2}\int_{V\times V} dx dy\{\omega(a^*(x)a(x+y)) - \omega(a^*(x))\omega(a(x+y))\} = 0 \qquad (2.33)$$

If $\rho_0 = 0$, then $c = 0$ or all terms vanish. However, if $\rho_0 > 0$, then also $c = \lim_V \omega(\frac{a_0}{\sqrt{V}}) = \omega(a(0)) \neq 0$, which means that the state can not be gauge invariant. The equation (2.33) expresses the property of the boson state ω of *showing off-diagonal long range order* [129].

The reader realizes that these explicit formulae, together with their interpretations, hold for the one- and two-point functions of any ergodic boson state independent of the fact that it is quasi-free or not quasi-free. Also the physical interpretations following below remain generally valid. These formulae hold for all homogeneous states of all homogeneous boson systems.

In particular one has the following physical picture. Looking at the definition formula of ρ_0, it is clear that it has to interpreted as the *zero-mode* or $(k=0)$-*density of particles* of the state ω. In particular, if for some boson system one can show that a ground or equilibrium state ω has the property $\rho_0 > 0$, then one can say that the state shows a macroscopic occupation of particles in the zero-mode. For such a state, the number of particles in the zero mode in a finite volume V has to increase proportionally with the volume (see Eq. (2.32)). Hence if

$$\rho_0 = \lim_V \frac{\omega(a_0^* a_0)}{V} > 0$$

one speaks about the occurrence of *Bose-Einstein condensation*, in short denoted by BEC, in that state and for the zero-mode. The value of ρ_0 itself is called the zero-mode *condensate density*.

Since by definition $\rho = \lim_V \omega(N_V/V)$, this quantity is called the *total density of particles* for the state ω. Since for all homogeneous states trivially holds that $\omega(a_k) = 0$ if $k \neq 0$, and because of Eq. (2.33) one gets for each ergodic state for which the thermodynamic limit $(V \to \infty)$ exists the formula

$$\omega\left(\frac{N_V}{V}\right) = \frac{1}{V}\sum_k \omega(a_k^* a_k) - \frac{1}{V}\omega(a_0^* a_0) + \frac{1}{V}\omega(a_0^* a_0) = \omega\left(\frac{N_V}{V}\right)_t + \frac{1}{V}\omega(a_0^* a_0)$$

One obtains for all ergodic states for which the total density ρ is finite, the following universal relation:

$$\rho = \rho_0 + \rho_c \tag{2.34}$$

It expresses that the total density for the state is the sum of the zero-mode *condensate density* ρ_0 and the density ρ_c of all excited $(k \neq 0)$ particles. The latter one is called the *critical density*, in the case that there is a non-trivial condensate $(\rho_0 > 0)$. The critical density is explicitly given in terms of the two-point operator $r(p)$ by the formula

$$\rho_c = \int dk\,(r(k) - 1) \tag{2.35}$$

In general the density relation Eq. (2.34) is of vital importance in the study of Bose-Einstein condensation for solvable as well as for non-solvable fully interacting boson models.

Sofar we considered only the one and two-point functions of an ergodic state. As is clear from the general definition Eq. (2.12), in order to fix a state completely one has to know all truncated n-point functions for all n=1,2,3,..... It is always important to keep in mind that they should satisfy the necessary and sufficient positivity conditions implied by the positivity of the state. We do not enter into an explicit discussion about these positivity properties. It is a straightforward but a technically annoying matter to write these properties out in their explicit form in terms of the correlation functions. We just repeat the remark that the ergodicity of the state implies that the higher order truncated n-point functions with $n > 2$, can be described, exactly as we did with the two-point functions, by multiplication operators in n-1 variables, and that this approach is useful.

In the case of open boson systems, condensation in excited modes can also occur. This means that one has the possibility of macroscopic occupation of a non-zero mode, say $q \neq 0$, for some state ω. This is expressed by the formula:

$$\rho_q = \lim_V \frac{\omega(a_q^* a_q)}{V} > 0$$

describing a macroscopic occupation of the q-mode. One speaks of the occurrence of q-condensation, which is a fair form of Bose-Einstein condensation. In Chapter Eq. (4) we discuss a simple model showing q-condensation.

So far we considered only the situation of fully space translation invariant states, i.e. states invariant for the full space translation group $\mathbb{R}^d$. However the situation of the space translation invariance with respect to a sublattice G of $\mathbb{R}^d$ is also relevant as will become clear in the applications below. In that case, consider a state ω which is invariant under the translation group $G = |a| \mathbb{Z} \times \mathbb{R}^{d-1}$, i.e. the continuous translations in $d - 1$ directions and the periodic lattice translations in one direction given by a vector of the form $a = |a| e$. It has a period of length $|a|$ and a direction along the unit vector e and it defines a corresponding momentum variable $q = e(2\pi/|a|)$. For each $x \in \mathbb{R}^d$ one can write $x = (ye, x_\perp)$, where $x_\perp$ is the x-component orthogonal to e and ye the e-component. For any such a G-invariant state $\widetilde{\omega}$ it could be meaningful to talk about the q-condensate density $\rho_q = \lim_V \widetilde{\omega}(\frac{a_q^* a_q}{V})$. Remark that one gets a full homogeneous state ω by integration over the period which is given by, for any observable X,

$$\omega(X) = \frac{1}{|a|} \int_0^{|a|} dy \, \widetilde{\omega} \circ \tau_y(X) \tag{2.36}$$

and a q-condensate density of the form

$$\rho_q = \lim_V \omega(\frac{a_q^* a_q}{V}) = \frac{1}{|a|} \int_0^{|a|} dy \, \widetilde{\omega}(a^*(y,0)a(y,0)) \tag{2.37}$$

Of course this construction extends to more dimensions up to all space directions.

All these definitions and physical interpretations about the one and two-point functions are of central importance in the language of physicists talking about fully interacting, respectively non-interacting boson systems studied with all the common techniques and concepts in many-body boson systems used since decades [81, 13, 79, 80]. As such they will remain of prime relevance in the rest of this monograph and in the future.

3

Equilibrium States

3.1 Variational Principle

It is well known that an equilibrium state at inverse temperature $\beta = \frac{1}{kT}$ can be determined by the so-called variational principle of statistical mechanics. Let k be the Boltzmann constant and T the absolute temperature of a homogeneous boson system determined by the local Hamiltonians H_V, with one Hamiltonian for each finite volume V. The principle is defined as follows: Consider the real map f, called the *grand canonical free energy density functional*, defined on the set of homogeneous or periodic states by the following. For any state ω of the system, f is defined by

$$f : \omega \to f(\omega) = \lim_V \frac{1}{V} (\beta \omega (H_V - \mu N_V) - S(\omega_V)) \tag{3.1}$$

where μ is the chemical potential, $N_V = \int_V dx\, a^*(x)a(x)$ the observable standing for the number of particles, and $S(\omega_V)$ the entropy of the restriction of the state ω to the finite volume V of $\mathbb{R}^n$. We indicate by ω_V this restriction of the state ω to the algebra $\mathfrak{A}_V$ of observables measurable within the volume V. This means that the set $\mathfrak{A}_V$ is generated by all creation and annihilation operators $a^{\#}(f)$ with test functions f having their support in V.

For simplicity we limit ourselves to those homogeneous states ω which are locally determined by a density matrix. In this case for each volume V, there exists a density matrix σ_V (i.e. $1 \geq \sigma_V \geq 0$, $tr\,\sigma_V = 1$) acting on the Fock vector space such that $\omega_V(A) = tr\,\sigma_V A$ for all A in $\mathfrak{A}_V$. States with this property are sometimes called *locally normal states*. We should realize immediately that this set of states contains all states of the Gibbs type. For any such locally normal state ω, we define the local entropy by $S(\omega_V) = -tr\,\sigma_V \ln \sigma_V$. For such homogeneous states the thermodynamic limit $V \to \infty$ in Eq. (3.1) can always be given a rigorous mathematical meaning, see [26, 149] for example.

The *grand canonical variational principle of statistical mechanics* expresses the following characterization of an equilibrium state:

A.F. Verbeure, *Many-Body Boson Systems*, Theoretical and Mathematical Physics, DOI 10.1007/978-0-85729-109-7_3, © Springer-Verlag London Limited 2011

Definition 3.1. *At an inverse temperature $\beta = 1/kT$, each homogeneous (or periodic) state ω_β, which minimizes the free energy density functional Eq. (3.1), is a grand canonical equilibrium state of the system, determined by the local Hamiltonians $\{H_V\}$ in the thermodynamic limit ($V \to \infty$) keeping a constant value for the particle density $\rho = \omega(N_V/V)$ for all volumes. In this limiting procedure, the chemical potential plays the role of a Lagrange multiplier with respect to conservation of particle density.*

This definition can equivalently be expressed by the following: The free energy density $f(\omega_\beta)$ of the system in any equilibrium state ω_β is bounded by the free energy density $f(\omega)$ in any other arbitrary homogeneous (or periodic) state ω of the system; or expressed mathematically, the equilibrium state ω_β satisfies the inequality

$$f(\omega_\beta) \leqq f(\omega) \tag{3.2}$$

It is proved for a large number of systems and for many different types (e.g., different types of the volumes V shapes) of thermodynamic limits, that the minimum of the free energy density functional is indeed reached [149] for one or more particular states. Hence this variational principle is a meaningful method to define equilibrium states. We stress that the infimum given by $f(\omega_\beta)$, which is equal to what in physics is called simply the free energy density of the system.

For any equilibrium state of each system, Criterion Eq. (3.1) can also be expressed as the equilibrium states that possess all the characteristics of *thermodynamic stability*. All equilibrium states yield the same lowest value of the free energy functional and all have the same free energy density. Any other homogeneous state, sometimes considered as a perturbed equilibrium state, yields a larger free energy density. Many more different notions of thermodynamic stability together with their own physical interpretation, are introduced and discussed in the literature, see [148, 151] for example. Here we stick to the best known one, that which was formulated above.

By taking the appropriate limit β tending to infinity (or T tending to zero) in the formulation of the variational principle Eq. (3.1), we immediately obtain the corresponding criterion characterizing the ground ($T = 0$) states of the system.

Definition 3.2. *A state ω_0 is a ground state of the system H_V if for any homogeneous state ω,*

$$\lim_V \omega\left(\frac{H_V - \mu N_V}{V}\right) \geq \lim_V \omega_0\left(\frac{H_V - \mu N_V}{V}\right) \tag{3.3}$$

It is clear that this criterion characterizes the ground states as the states of lowest energy density. Furthermore, to guarantee solutions to this criterion we assume that all systems under consideration, except for a couple of models, are mechanically stable, which leads to the following definition:

Definition 3.3. *All local Hamiltonians H_V satisfy the mechanical stability condition which is expressed by: For all large but finite volumes V, there exist real numbers a and b with $b \geq 0$ such that*

$$H_V \geq b\frac{N_V^2}{V} - aN_V \tag{3.4}$$

In this case we call the system $\{H_V\}_V$ stable. The system is called super-stable if in addition $b > 0$.

From the practical point of view, looking for the equilibrium states of any quantum system and in particular of a boson system, it means looking for the solutions of the variational principle Eq. (3.1). For boson systems it is a matter of deriving from this principle all n-point correlation functions of one, or sometimes more, equilibrium states ω_β. In particular, when looking for ergodic solutions we must realize that the free energy density $f(\omega)$ of any state ω is expressed by the one-point function $c = \sqrt{\rho_0}e^{i\varphi}$; the two-point truncated correlation functions $r(p), s(p)$, or $t(p)$ satisfying the positivity conditions; and all the remaining higher-order ($n > 2$)-point truncated correlation functions of the states. Without going into all details here, looking for a solution ω_β means solving a variational principle with an infinity of a priori unknown variables, which we should realize is in general an infinitely difficult problem.

Looking closer at this problem, one can consider first the variations with respect to the one-point functions. Note that there is no a priori positivity condition on the phase parameter φ, the argument of the parameter c, the one-point function. It is a free parameter, for which we can straightforwardly derive its corresponding Euler equation and seek its solutions. Then we can perform the variation with respect to the square root of the condensate density $\sqrt{\rho_0} = |c|$. This is already a more delicate affair because this is not a free parameter. It is not immediately clear within which range this parameter can vary because of the total ρ-density constraint Eq. (2.34). This variational operation is extensively studied and worked out further in section (4.3.3) of Chapter (4), where we discuss and derive the condensate equations. Later on we learn that the latter ones can be considered as the quantum Euler equations for all variations in the order parameters of this quantum variational problem. In particular this holds for the square root of the condensate density. For explicit treatments of the derivation of all other so-called quantum Euler equations to be satisfied by all other ($n \geq 2$)-point functions, we refer mainly to the multiple model applications in Chapter (4).

As a matter of a motivation for the infinite-volume variational principle Eq. (3.1) to be a good definition of equilibrium states in the thermodynamic limit situation, it is instructive to analyze in somewhat more detail the variational principle for finite volume systems. If the system is (super-)stable, then for all finite $\beta > 0$ the partition function $tr \exp(-\beta(H_V - \mu N_V)) < \infty$ is finite for each finite volume V such that the variational principle can be formulated for each finite volume. The solution of the principle is unique and yields the grand canonical Gibbs state as a unique equilibrium state solution. The explicit proof of this statement may be inspiring, as we now offer:

We use the simplified notation $H(\mu) = H_V - \mu N_V$. The corresponding Gibbs state is given by: For any observable A, the Gibbs state or the Gibbs equilibrium expectation-valued map ω_β is given by

$$\omega_\beta(A) = \frac{tr\, e^{-\beta H(\mu)}A}{tr\, e^{-\beta H(\mu)}} \tag{3.5}$$

We repeat that a *normal state* of a boson system is a state ω determined by a density matrix ρ ($tr\rho = 1; 0 \leq \rho \leq 1$) acting on the Fock Hilbert space $\mathfrak{F}$ such that

$$\omega(A) = tr\rho A \tag{3.6}$$

Clearly the Gibbs state is a normal state with density matrix

$$\rho_\beta = \frac{e^{-\beta H(\mu)}}{tre^{-\beta H(\mu)}}$$

The *free energy* of a system $H(\mu)$ is given by

$$F(\beta, \mu) = -\log tre^{-\beta H(\mu)},$$

The *entropy* of a normal state, determined by the density matrix ρ, is defined as usual by the expression $S(\rho) = -tr\rho \ln \rho$ and the *internal energy* by the expectation value $E(\rho) \equiv E(\omega) \equiv \omega(H(\mu))$. We define, as above for the densities, the *free energy functional* now on the set of all density matrices ρ acting on the Fock Hilbert space. Hence for all normal states the free energy functional $F(\rho)$ for the density matrix ρ, takes the form

$$F(\rho) = \beta E(\rho) - S(\rho).$$

The basic variational principle of statistical mechanics for finite systems is now formulated as the following theorem:

Theorem 3.4. *The free energy of the system satisfies*

$$F(\beta, \mu) = \inf_\rho F(\rho) = F(\rho_\beta)$$

where the inf is taken over all density matrices and where ρ_β, is the density matrix of the Gibbs state. In other words, the Gibbs state is the unique solution of this finite system variational principle.

Proof. First check that $F(\beta, \mu) = F(\rho_\beta)$ by an explicit and simple computation. Hence the Gibbs state yields the free energy of the system as expected.

Furthermore, compute the expression

$$F(\rho) - F(\rho_\beta) = tr(\rho \ln \rho - \rho \ln \rho_\beta)$$

Let $(f_i)_i$ and $(e_j)_j$ be orthonormal bases which diagonalize the density matrices ρ, respectively ρ_β, with the eigenvalues ρ_i and $\rho_{\beta,j}$, then

$$F(\rho) - F(\rho_\beta) = \sum_i \sum_j |(f_j, e_i)|^2 (\rho_i \ln \rho_i - \rho_i \ln \rho_{\beta,j})$$

Using the strict convexity of the function, $f(x) = x \ln x$ for $x \geq 0$, yielding

$$(y - x)f'(x) \leq f(y) - f(x)$$

with the equality sign if and only if $x = y$. Use the fact that ρ and ρ_β are density matrices, in particular satisfying $tr\rho = tr\rho_\beta = 1$, we find for all density matrices ρ: $F(\rho) \geq F(\rho_\beta)$ with the equality sign if and only if $\rho = \rho_\beta$, therefore proving the theorem.

This theorem is nothing more than a formalization of what is standard knowledge within the physics community. Here it is mentioned as an argument in favor of the definition of the formulated variational principle of statistical mechanics for homogeneous systems in the thermodynamic limit Eq. (3.1). We see that the main difference between the criteria Eq. (3.4) and Eq. (3.1), is that in the second one the thermodynamic functions are replaced by their densities to make all ingredients of the formulation mathematically and physically meaningful. It is clear from Eq. (3.4) that the thermodynamic limit of any Gibbs state satisfies the principle Eq. (3.1). It is also important to realize the following difference between the two principles Eq. (3.1) and Eq. (3.4): As proved above, the finite volume principle Eq. (3.4) has only one solution, namely the Gibbs state. Hence its set of solutions is automatically a singleton, which evidently cannot be decomposed into a non-trivial convex combination of two or more other different solutions or equilibrium states. Clearly this finite volume principle is only able to describe one-phase physics situations.

Therefore in order to work with multiple phase physical systems, phase transitions and all that, we must apply the variational principle Eq. (3.1) formulated in the thermodynamic limit. The challenge accompanying this principle is to derive from it all its solutions. We denote all of them with the simple general notation ω_β. What does all this mean in practice? There is only one way of doing and this is to derive from Eq. (3.1) all the correlation functions Eq. (2.7) of the unique equilibrium state or of all the possible equilibrium states. In the following chapters we find a number of practical hints and ways of proceeding to that goal for a number of boson system models, as well as for the general two-body interacting particles model. In particular a general practical tool in this search for a derivation of properties of the equilibrium states is to consider the condensate equations. They represent some of the quantum Euler equations of the variational principle. For instance one of these equations turns out to be an explicit closed equation for the condensate density. Finding a non-trivial solution of this equation for the condensate density is akin to proving the existence of equilibrium states showing Bose-Einstein condensation and in due case other types of phase transitions. An extensive and detailed account of the condensate equations is found in the next chapter.

3.2 Energy-Entropy Balance Criterion

A characterization of equilibrium states, even more generally applicable than the variational principle of the previous section, is now given in terms of the energy-entropy balance correlation inequalities. These conditions are more general than the variational principle because they are also applicable as equilibrium criteria in the case of non-homogeneous systems. Moreover they give an interesting physical interpretation of the nature of an equilibrium state. Finally these correlation inequalities turn out to be highly practical tools for many applications. As this characterization of equilibrium is less known in the physics literature, we develop first a detailed discussion of this principle. It might help as an intellectual motivation in order to use it as a characterization of equilibrium.

As a matter of introducing this subject, we consider again the case of density matrix states for finite systems with a Hamiltonian $H(\mu)$ as we did in the previous subsection in order to argue the variational principle Eq. (3.1). We start with the following property:

Theorem 3.5. *The necessary and sufficient condition in order that a state ω is the Gibbs state ω_β, is that it satisfies the following correlation inequalities, called from now on the energy-entropy balance correlation inequalities: For any observable A,*

$$\beta \omega(A^*[H(\mu),A]) \geq \omega(A^*A) \ln \frac{\omega(A^*A)}{\omega(AA^*)}$$

where the function $f(u,v) = u \ln \frac{u}{v}$ is well defined for all real $u,v > 0$ and where $f(u,v) = 0$ if $u = v = 0$.

Proof. Consider the spectral resolution of the Hamiltonian $H(\mu)$: Let $(e_i)_i$ be the orthonormal basis that diagonalizes the Hamiltonian and let $E_{i,j}$ be the matrix units or partial isometries (in Dirac notation) $E_{i,j} = |e_i \rangle\langle e_j|$, mapping the vector e_j onto e_i. Then $H(\mu) = \sum_i \varepsilon_i E_{i,i}$, with $(\varepsilon_i)_i$ the spectral values of $H(\mu)$.

Take first $\omega = \omega_\beta$, that is, ω is the Gibbs state. Using the cyclic permutation property of the trace $tr CD = tr DC$, we get

$$\frac{\omega_\beta(AA^*)}{\omega_\beta(A^*A)} = \frac{\omega_\beta(A^* e^{-\beta H(\mu)} A e^{\beta H(\mu)})}{\omega_\beta(A^*A)} = \sum_{i,j} \frac{\omega_\beta(A^* E_{i,i} A E_{j,j})}{\omega_\beta(A^*A)} e^{-\beta(\varepsilon_i - \varepsilon_j)}$$

Note that $\omega_\beta(A^* E_{i,i} A E_{j,j}) \geq 0$ and that $\sum_{i,j} \omega_\beta(A^* E_{i,i} A E_{j,j}) = \omega_\beta(A^*A)$. Using the convexity of the exponential function we obtain the following inequalities for each observable A:

$$\frac{\omega_\beta(AA^*)}{\omega_\beta(A^*A)} \geq \exp\{-\beta \sum_{i,j} (\varepsilon_i - \varepsilon_j) \frac{\omega_\beta(A^* E_{i,i} A E_{j,j})}{\omega_\beta(A^*A)}\} = \exp\{-\frac{\beta \omega_\beta(A^*[H(\mu),A])}{\omega_\beta(A^*A)}\}$$

proving that the Gibbs state satisfies the correlation inequalities of the theorem.

Now we prove the converse. Therefore we start with an arbitrary normal state $\omega(.) = tr \rho.$, satisfying the correlation inequalities of the theorem for all observables A. We have to prove that this state coincides with the Gibbs state.

For the proof, take first $A^* = A$. Then the inequalities yield $\omega(A[H(\mu),A]) \geq 0$ and hence $\overline{\omega(A[H(\mu),A])} = \omega(A[H(\mu),A])$ expressing that it is a real number. But this equality means also that $\omega([H(\mu),A^2]) = 0$.

As each operator A can be written as a linear combination of at most four positive operators, it follows that $\omega([H(\mu),A]) = 0$ holds for each observable A. We should realize that this result already means that the state ω is time invariant for the dynamics defined by the given system Hamiltonian $H(\mu)$. Using again the cyclic permutation property under the trace, we get $[\rho, H(\mu)] = 0$, or ρ and $H(\mu)$ commute, and therefore they can be diagonalized with respect to the same basis. In particular we have $\rho = \sum \rho_i E_{i,i}$ with all $\rho_i \geq 0$. Substitute now $A = E_{l,k}$ in the correlation inequalities, we obtain

$$\beta(\varepsilon_l - \varepsilon_k)\omega(E_{k,k}) \geq \omega(E_{k,k}) \ln \frac{\omega(E_{k,k})}{\omega(E_{l,l})}$$

as well as the equivalent inequality with l and k interchanged. Suppose that $\omega(E_{l,l}) = 0$. It then follows that $\omega(E_{k,k}) = 0$ for all k. This is impossible because $\sum \omega(E_{k,k}) = 1$, which is a consequence of the normalization of the state. Hence for all l: $\omega(E_{l,l}) > 0$. Therefore we get

$$\beta(\varepsilon_k - \varepsilon_l) = \ln \frac{\omega(E_{l,l})}{\omega(E_{k,k})}$$

$$\omega(E_{l,l})e^{\beta \varepsilon_l} = \omega(E_{k,k})e^{\beta \varepsilon_k} = \lambda$$

with λ a constant independent of the indices k and l. From the normalization of the state again follows that $\lambda = 1/tr e^{-\beta H(\mu)}$. Hence $\rho_k = \omega(E_{k,k}) = e^{-\beta \varepsilon_k}/tr e^{-\beta H(\mu)}$ or $\rho = \sum_k \frac{e^{-\beta \varepsilon_k}E_{k,k}}{tr e^{-\beta H(\mu)}} = \frac{e^{-\beta H(\mu)}}{tr e^{-\beta H(\mu)}}$, is equal to the Gibbs density matrix and ω is the Gibbs state.

This theorem, together with the previous one Eq. (3.4), show that the variational principle, as well as the energy-entropy balance correlation inequalities, yield equivalent characterizations of the canonical(grand canonical) Gibbs states for finite volume systems. Moreover both equivalent formulations yield for each of them an interesting physical characterization of an equilibrium state, or of equilibrium in general. We will return to that point later.

In any case, on the basis of the theorem Eq. (3.5) it is as quite possible to define the equilibrium state of an infinitely extended system, a system in the thermodynamic limit ($V \rightarrow \infty$), as a solution of the correlation inequalities Eq. (3.5). For these reasons we make the following formal definition of an equilibrium state for infinitely extended systems:

Definition 3.6. *The Energy-Entropy-Balance(EEB) criterion yields the following criterion for equilibrium states in the thermodynamic limit: Each state ω_β of the system satisfying, for each observable A in the domain of the commutator $\lim_V [H_V - \mu N_V, .]$, for each fixed inverse temperature β and chemical potential μ, the inequalities*

$$\beta \lim_V \omega_\beta (A^*[H_V - \mu N_V, A]) \geq \omega_\beta(A^*A) \ln \frac{\omega_\beta(A^*A)}{\omega_\beta(AA^*)} \tag{3.7}$$

is an equilibrium state of the system at the inverse temperature β and chemical potential μ.

Again as for the variational principle, taking the limit $\beta \rightarrow \infty$ of the EEB criterion yields the following defining criterion for the ground states:

Definition 3.7. *Any state ω_0 satisfying $\lim_V \omega_0(A^*[H_V - \mu N_V, A]) \geq 0$ for any observable A, is a ground state of the boson system defined by the Hamiltonians H_V and particle density determined by the chemical potential μ.*

For homogeneous systems it is easy to check that this ground state criterion is equivalent to the one based on the variational principle. As before, in the following we work in the grand canonical ensemble and for notational convenience we continue with the notation $H_V(\mu) = H_V - \mu N_V$.

It is clear that as well the EEB criterion as the variational principle get both a formulation characterizing the (grand-)canonical equilibrium states.

From a practical point of view, to derive and know the equilibrium states on the basis of the EEB criterion, we must obtain again, as in the case of the variational principle, all n-point correlation functions of the equilibrium state(s), but now from the EEB-inequalities Eq. (3.7). The hint to proceed with this task is clearly to choose in a clever way several local observables A, to substitute them in the inequalities, and to analyze the outcomes. To gain a view on a number of especially interesting hints for the best choice among the observables, we refer again to the next chapter, where for several models the equilibrium solutions are also explicitly computed on the basis of the EEB criterion. In particular for the computation of the one-point function we refer again to Eq. (4.3.3) for the derivation of the corresponding condensate equations.

We should note that the variational principle and EEB criteria all hold as equilibrium criteria for all quantum systems, not necessarily just arbitrary boson systems. Before proceeding to the next chapter, we derive from the EEB criterion a number of generally valid properties of equilibrium states for any quantum system as well as, of course, any boson system. As a first result following from this criterion, the stationarity or the time invariance of all equilibrium states is derived. Although this property may come over as an obvious physical statement, its derivation is not sufficiently visible in the physics literature. We formulate it as the following theorem:

Theorem 3.8. *Let ω_β be a state, a solution of the EEB criterion Eq. (3.7). Then for each observable X,*

$$\lim_V \omega_\beta([H_V(\mu), X]) = 0 \qquad (3.8)$$

which expresses the time invariance or the stationarity of any equilibrium state ω_β.

Proof. We first substitute an arbitrary self-adjoint observable $X = X^*$ into the inequality Eq. (3.7), obtaining

$$\lim_V \omega_\beta(X[H_V(\mu), X]) \geq 0$$

which demonstrates that the left hand side of this inequality is a real number. Therefore

$$\lim_V \omega_\beta(X[H_V(\mu), X]) = \lim_V \overline{\omega_\beta(X[H_V(\mu), X])}$$
$$= \lim_V \omega_\beta((X[H_V(\mu), X])^*)$$
$$= -\lim_V \omega_\beta([H_V(\mu), X]X)$$

Hence

$$0 = \lim_V \omega_\beta \left(X[H_V(\mu), X] \right) + \lim_V \omega_\beta \left([H_V(\mu), X]X \right) = \lim_V \omega_\beta \left([H_V(\mu), X^2] \right)$$

and the theorem is proved for the positive operators of the form $Y = X^2$. As this relation is linear in Y, and as every operator can be written as a linear combination of four positive operators, it follows that the relation Eq. (3.8) holds for any arbitrary observable X.

The stationarity of the equilibrium states as stated in this theorem is sometimes referred to as its infinitesimal form. In Chapter (5) we discuss the dynamics of Bose systems and relate this formulation to the more common property of stationarity. Of course the stationarity can as well be derived directly from the variational principle Eq. (3.1).

For a more mathematical introduction of the EEB criterion together with a study of its equivalence with other correlation inequalities and other characterizations of equilibrium we refer to [52] and for related notions to [148, 151].

The physical interpretation of this EEB criterion can also be be worked out in greater detail. For more information about this point we refer to [53] for the following interpretation of the criterion: The left hand side of the inequalities Eq. (3.7) represents the change of energy in the equilibrium state under a dissipative perturbation, which is locally generated by the observable X. In fact, it is the time derivative of the energy density under a dissipative dynamics which we discuss in Chapter (5) and (7.2)). The right hand side, or the lower bound, represents the change in entropy of the equilibrium state under the same dissipative perturbation of the equilibrium state. The EEB criterion Eq. (3.7) tells us that an equilibrium state is completely characterized by the fact that if an equilibrium state gets perturbed its energy increase is always majoring its entropy increase. A state not satisfying this criterion and therefore showing wild changes of the entropy not bounded by the energy changes cannot be an equilibrium state. This is the sense in which the EEB correlation inequalities express the property of thermodynamic stability of each equilibrium state.

For a more explicit study about the equivalence of the equilibrium EEB criteria Eq. (3.7) and the variational principle Eq. (3.1) for homogeneous systems in the thermodynamic limit we refer to [53] and [26]. In fact both criteria express the same thermodynamic stability. The EEB criterion has to be considered as a differential form of the variational principle comparable to the position of the Euler equations in standard variational analysis. We should not be surprised by the inequalities, instead of equalities, because we can prove that the inequalities are equivalent to Euler equations (see [53] for example). The latter are however practically less manageable in application. It will become clear later that the inequalities are handy tools for the study of equilibrium properties.

We need to realize that the EEB- criterion for equilibrium states is given by inequalities expressed in terms of the correlation functions which determine completely the equilibrium state.

A number of other correlation inequalities known in the literature have shown to be useful tools in theoretical and mathematical physics as well. In particular we mention a well-known correlation inequality, namely the so-called *Bogoliubov in-*

equality [120, 77, 25] for Gibbs states, which turned out to be an interesting tool in disproving the existence of spontaneous symmetry breaking (see Chapter (4)) or of phase transitions in boson and other quantum systems. This inequality provides an interesting upper bound on the quantum fluctuations. The question can be raised about its relation to the above EEB-correlation inequalities. It should be clear that as the EEB criterion is a full characterization of the equilibrium states, all other correlation inequalities for equilibrium states should follow from it. As an illustration of this fact, we derive now the well known Bogoliubov inequality and another useful inequality, called the *double commutator inequality*, directly from the EEB-correlation inequalities.

Theorem 3.9. *Let ω_β be any equilibrium state satisfying the EEB criterion then for each observable X holds the Bogoliubov inequality:*

$$\lim_V \beta \omega_\beta([X^*,[H_V(\mu),X]])\frac{1}{2}\omega_\beta(X^*X+XX^*) \geq |\omega_\beta([X,X^*])|^2 \qquad (3.9)$$

implying immediately the double commutator inequality

$$\lim_V \omega_\beta([X^*,[H_V(\mu),X]]) \geq 0 \qquad (3.10)$$

Proof. The EEB-inequality Eq. (3.7) is an inequality on the real numbers, hence each side of the inequality should be real. In particular

$$\lim_V \omega_\beta(X[H_V(\mu),X^*]) = \overline{\lim_V \omega_\beta(X[H_V(\mu),X^*])} = -\lim_V \omega_\beta([H_V(\mu),X]X^*)$$

Therefore again from Eq. (3.7), considering the inequalities for X and for X^*, and after adding both of them, we get

$$\beta \lim_V \omega_\beta([X^*,[H_V(\mu),X]]) \geq \omega_\beta([X^*,X]) \ln \frac{\omega_\beta(X^*X)}{\omega_\beta(XX^*)}$$

As for all $a,b > 0$ we get $(a-b)\ln(a/b) \geq 0$; the second inequality of the theorem follows immediately. Furthermore, for all $a \geq b > 0$ (the other case $b \geq a > 0$ is similar), $\ln(a/b) = \int_b^a dx f(x)$, with $f(x) = 1/x$ being a convex function on the interval $[b,a]$. Therefore $f(x) \geq g(x) = f'(\frac{a+b}{2})(x - \frac{a+b}{2}) + f(\frac{a+b}{2})$ and hence $\int_b^a dx f(x) \geq \int_b^a dx g(x) \equiv 2\frac{a-b}{a+b}$, therefore proving the Bogoliubov inequality.

We should realize that the set of Bogoliubov inequalities for arbitrary observables does not constitute a full criterion for equilibrium states comparable with the EEB criterion. In other words, the Bogoliubov inequalities do not imply the EEB-inequalities. This follows from the simple fact that the *central state*, which we denote by ω_∞, always satisfies the Bogoliubov inequality. The central state is defined by the property that, for each pair of observables X, Y, $\omega_\infty(XY) = \omega_\infty(YX)$, that is under the state two operators always commute. Also, the central state is obtained from a Gibbs state after having taken the limit of the temperature going to infinity (or $\beta = 0$). Therefore physically the central state can be referred to as the infinite temperature state. This fits with the popular wisdom of quantum physics being a low temperature phenomenon, classical physics a high temperature one.

Finally we note that in the literature often the following inequality holds: For each pair of observables X, Y and for each equilibrium state ω_β holds the inequality

$$\lim_V \beta \, \omega_\beta([X^*, [H_V(\mu), X]]) \frac{1}{2} \omega_\beta(Y^*Y + YY^*) \geq |\omega_\beta([X, Y^*])|^2 \qquad (3.11)$$

This inequality is referred to as the *Bogoliubov inequality*. This inequality, with two possibly different observables X and Y, looks more general than the above Eq. (3.9) formulated in the theorem. It is however easy to show that both are just two different formulations of the same inequality. The proof of this equivalence is obtained by an easy algebraic argument and is left as an exercise. This proof is also found in [52].

3.3 Variational Principle for Solvable Models

In this section we describe explicitly the most general form of the variational principle when one is faced with homogenous solvable models. Before we define explicitly what we mean to have a solvable model, we introduce a new notion. Its relevance will become clear in the derivation of the variation principle for solvable boson models, which is the main contribution of this section.

Definition 3.10. *Let ω be an arbitrary boson state. The two-point truncated functions of the state define two operators R and S, exactly as explained in Eq. (2.3). But in turn this couple of operators defines also a quasi-free state, which we denoted $\omega_{(R,S)}$. The latter state is always essentially different from the given state except if the latter is already a qf-state. We call the state $\omega_{(R,S)}$ the quasi-free state associated to or induced by the given (general) state ω.*

As an explicit and useful example of a state and its associated qf-state, we consider a normal boson state ω_ρ, with a density matrix ρ acting on the Fock Hilbert space $\mathfrak{F}$, explicitly $\omega_\rho(.) = tr\rho(.)$. In general this is not a quasi-free state. We construct explicitly its associated qf-state $\omega_{(R,0)}$.

Let us consider an orthogonal normal basis $\{f_m\}$ of the test function space $\mathscr{S}$ diagonalizing the density matrix ρ and defining the non-negative real numbers n_k, by the relations $n_k \delta_{k,l} = tr\rho\, a_k^* a_l$ with the notation $a_l = a(f_l)$. Consider now the operator, sometimes called the effective Hamiltonian of the state (see also Chapter Eq. (5)), $H = \sum_k \varepsilon_k a_k^* a_k$ with the energy values $\varepsilon_k = \ln((n_k + 1)/n_k)$ and consider also the new density matrix $\sigma = e^{-H}/tr\,e^{-H}$. Let us define the normal state $\omega_\sigma(.) = tr\sigma(.)$, with density matrix σ. We then compute

$$tr\,\sigma\, a_k^* a_l = tr\rho\, a_k^* a_l = \delta_{k,l}\, n_k \qquad (3.12)$$

which expresses the property that the two-point functions of the states ω_ρ and ω_σ do coincide.

Note that ω_σ is a quasi-free state. In the chosen basis, in fact, it is the Gibbs state for the (free particles) Hamiltonian H. It is identified to be equal to the qf-state $\omega_\sigma = \omega_{(R,0)}$ with R the diagonal matrix $((n_k+1)\delta_{k,l})_{k,l}$.

It is a satisfying exercise to verify that its truncated correlation functions of order larger than 2 all vanish. The state ω_σ is the associated qf-state induced by the given state ω_ρ.

We use this construction to derive an entropy inequality between the entropies of each normal state and its associated qf-state. For any normal state ω_ρ with density matrix ρ as above, the (von Neumann) entropy is defined as always by $S(\omega_\rho) = -tr\rho\ln\rho$.

Lemma 3.11. *We obtain the entropy inequality*

$$S(\omega_\rho) \leq S(\omega_{(R,S)}) \tag{3.13}$$

where $\omega_{(R,S)}$ is the quasi-free state associated to the state ω_ρ.

Proof. As above, using a Hilbert space basis diagonalizing the density matrix ρ of ω_ρ such that the couple (R,S) simplifies to the form $(R,0)$, we obtain

$$S(\omega_{(R,0)}) - S(\omega) = tr\rho\ln\rho - tr\sigma\ln\sigma \tag{3.14}$$

Using Klein's convexity inequality [26], Lemma 6.2.21, that is, simply using the convexity of the function $f(x) = x\ln x$ (see the proof of Eq. (3.4)), we obtain

$$tr(\rho\ln\rho - \sigma\ln\sigma) \geq tr(\rho - \sigma)\ln\sigma \tag{3.15}$$

where $\ln\sigma = -\sum \varepsilon_k a_k^* a_k - \ln tr(\exp -H)$ and hence

$$S(\omega_\sigma) - S(\omega\rho) \geq -\sum \varepsilon_k(tr\rho\,a_k^* a_k - tr\sigma\,a_k^* a_k) = 0$$

because the states ρ and σ have the same two-point functions. This proves the inequality Eq. (3.13).

This inequality is a mathematical expression with the following physical interpretation: As is commonly understood, the von Neumann formula for the entropy of a state expresses the degree of disorder in the state. With this in mind we need to remember that ω_ρ is a state with more non-trivial correlations than its associated qf-state $\omega_{(R,S)}$ because infinitely many truncated correlation functions are potentially set equal to zero. Therefore we have to expect that the entropy of the state ω_ρ is smaller than or equal to the entropy of its associated qf-state ω_σ. That is exactly what we proved and expressed in the preceding Lemma.

In Eq. (3.1), we formulated the general variational principle of statistical mechanics valuable for an arbitrary boson system H_V in the thermodynamic limit. We denote by ω_V the restriction of the state ω to the algebra of observables measurable within the finite volume V. Without restriction of generality we can assume that ω_V is a normal state, or equivalently that it is determined by the density matrix ρ_V in

$\omega_V(.) = tr\rho_V(.)$. Of course the main reason for this assumption is that we are basically interested in equilibrium states that are thermodynamic limits of local Gibbs states, which are all normal states by construction. In other words, for our purposes we may assume that all states ω under consideration in the variational principle are locally normal. Therefore we can rewrite the principle Eq. (3.1) as follows:

Definition 3.12. *Any equilibrium state ω_β at inverse temperature $\beta = \frac{1}{kT}$ and chemical potential μ is a solution satisfying the variational principle*

$$f(\omega_\beta) = \inf_\omega f(\omega) = \inf_\omega \lim_{V\to\infty} (\beta\, tr\rho_V(H_V - \mu N_V) + tr\rho_V \ln\rho_V) \qquad (3.16)$$

where f is the free energy density functional Eq. (3.1) and where the minimum is taken over all homogeneous states.

In the rest of this section we concentrate our attention on the question of *solvable systems* or *solvable models*, which are defined by the local Hamiltonian $\{H_V\}_V$. The first question is: What is a solvable model?

Definition 3.13. *The model H_V is a solvable model if the energy (minus the chemical potential term) density functional for any ergodic state ω, which is given by*

$$e(\omega) \equiv \lim \omega(\frac{H_V - \mu N_V}{V}) = \lim \omega_V(\frac{H_V - \mu N_V}{V}) = \lim tr\rho_V(\frac{H_V - \mu N_V}{V}), \quad (3.17)$$

depends only on the one- and two-point truncated correlation functions of the state ω and does not depend on any of the $(n > 2)$-point functions. Equivalently this expression depends only on the one-point functions and on the operators (R,S) defining the ω-associated qf-state $\omega_{(R,S)}$. The one-point functions are again given by $\omega(a(f))$ and its complex conjugate. In other words the energy density of any solvable model for any ergodic state ω depends only on the quasi-free character of the state, that is, this energy density equals to the energy density for the associated qf-state of ω.

We know that for homogeneous states the one-point functions take the form $\omega(a(f)) = cf(p = 0)$, where c is a complex constant. Therefore it is reasonable to denote the most general qf-state with the extended notation $\omega_{(R,S,c)}$. In particular we denote also $\omega_{(R,S,0)} \equiv \omega_{(R,S)}$ whenever $c = 0$.

The energy density functional value in the state ω for a solvable model is therefore characterized by the functions (r,s) and the constant c, or if $\alpha(p) = \arg s(p)$ by $(r \geq 1, t \geq 0, \alpha, c)$. Hence the energy density $e(\omega)$ of any solvable model in any state ω has the following dependence on its state parameters: $e(\omega) \equiv e(r,t,\alpha,c)$.

Consider now the entropy term in the variational principle Eq. (3.16). Because the principle is restricted to locally normal states we can use the following property:

For normal states, that is, for density matrix states, as an immediate consequence of von Neumann's uniqueness theorem (see e.g. [131], Chapter 1 and 9) each canonical transformation τ is implemented by a unitary operator Eq. (7.3). This means that there exists a unitary operator U such that $\tau(A) = UAU^*$ for each observable A. Therefore $\omega_V(\tau(A)) = tr\rho_V UAU^* = trU^*\rho_V UA$ and the density matrix of the state $\omega_V \circ \tau$ is given by $U^*\rho_V U$. Because of the property $trU^*XU = trX$, the entropy functional is left invariant under any canonical transformation yielding $S(\omega_V \circ \tau) = S(\omega_V)$. Two immediate consequences are in order:

i) The entropy does not depend on the one-point functions (see Eq. (2.15) and thereafter). Therefore we continue with qf-states of the type $\omega_{(R,S)}$.

ii) From Lemma Eq. (2.6) it follows that for any such qf-state, $S(\omega_{(R,S)}) = S(\omega_{(\tilde{R},0)})$. This means that the problem of computing explicitly the entropy density is reduced to its computation for gauge invariant qf-states.

Furthermore in [47] the explicit formula of the entropy density for gauge invariant homogeneous qf-states in terms of function $r(p)$, which is well known in the physics literature, is rigorously proven. We obtain the following expression for the entropy density of any arbitrary qf-state $\omega_{(r,t,\alpha,c)}$:

$$\begin{aligned} s(r,t,\alpha,c) &= s(r,t,0,0) = s(\omega_{(R,S)}) = s(\omega_{(\tilde{R},0)}) \\ &= \lim_V \frac{S(\omega_{(\tilde{R}_V,0)})}{V} \\ &= \int dp\,(\tilde{r}(p)\ln\tilde{r}(p) - (\tilde{r}(p)-1)\ln(\tilde{r}(p)-1)) \end{aligned}$$
(3.18)

where $\tilde{r}(p)$ is given as a function of r and t in the Lemma Eq. (2.6).

Using all these results regarding the energy and the entropy densities, and in particular the inequality Eq. (3.13), we obtain, for any homogeneous state ω with associated qf-state determined by the parameters (r,s,α,c), the explicit expression of the free energy functional for any solvable model and the inequality

$$f(\omega) = \beta e(\omega) - s(\omega) \geq \beta e(r,t,\alpha,c) - s(r,t,\alpha,c)$$
(3.19)

Hence the inequality Eq. (3.13) proves essentially that, for solvable models, the minimum over all states in the general variational principle Eq. (3.16) reduces to a variational principle over the smaller set of the homogeneous quasi-free states. The latter set is a closed subset of the set of all ergodic states of the system.

Hence we proved the following theorem:

Theorem 3.14. *Let H_V be the local Hamiltonian of a solvable boson system. Then the variational principle of statistical mechanics for its equilibrium states Eq. (3.16) reduces to the variational principle over the set of all homogeneous quasi-free states, a property explicitly expressed by*

$$\begin{aligned} f(\omega_\beta) &= \inf_{\omega \in \Omega} f(\omega) \\ &= \inf_{r \geq 1, t \geq 0, \alpha, c} (\beta e(r,t,\alpha,c) - s(r,t,0,0)) \\ &= \inf_{r \geq 1, t \geq 0, \alpha, c} \{\beta e(r,t,\alpha,c) - \int dp\,\{\tilde{r}(p)\ln\tilde{r}(p) - (\tilde{r}(p)-1)\ln(\tilde{r}(p)-1)\}\} \end{aligned}$$

where $\tilde{r}(p)$ is given by Lemma Eq. (2.6) as a function of r and t. The solutions, namely the equilibrium states ω_β, are always quasi-free states.

This theorem teaches us that for solvable systems the search for equilibrium states is largely reduced to a problem of classical analysis. The problem of solving the quantum Euler equations, has been turned explicitly to a problem of variations on the following sets of functions, namely the sets of continuous non-negative functions $r-1$ and t, the real continuous functions α and the complex constants c.

About this explicit variational problem we mention that the parameter α is a free parameter yielding no constraints and we compute directly its corresponding Euler equation. The variational equation with respect to the parameter $|c| = \sqrt{\rho_0}$, the square root of the condensate density, is not as straightforward because the values of this parameter are, in the grand canonical version of the principle, constrained by the constant total density $\rho = \rho_0 + \rho_c$. We do not enter here into more details but we refer again to section Eq. (4.3.3) of Chapter Eq. (4), where we discuss the condensate equation.

We now consider the variational operation of the free energy functional on the set of non-negative continuous functions $r-1$ such that $\int dp\, (r(p) - 1) \leq \rho$, where ρ is again the given constant density. Finally, we perform the variation on the set of non-negative continuous functions t satisfying the constraint $\sqrt{r(p)(r(p) - 1)} \geq t(p) \geq 0$. This step finishes the main steps in the realization of the complete variational principle for any solvable boson system. We will explicitly solve a number of solvable models for their equilibrium states according to this variational principle in the next chapter.

Many of us may wonder whether for solvable models the search for equilibrium states could be simpler or not if we had used the energy-entropy criterion for equilibrium states. It is clear that the notion of a solvable model leads indeed also to essential simplifications when searching for solutions of the EEB criterion. In particular it will be clear from reading Chapters Eq. (4) and Eq. (5) that to obtain a complete solution the choice of the observables X in the EEB inequalities can be restricted to linear combinations of the creation and annihilation operators. Using only these operators in the EEB inequalities is sufficient to specify all correlation functions of all equilibrium states. The specific ways of proceeding and the arguments for these statements become explicit (and hopefully clear) in the applications. In any case both equilibrium criterions lead to essential simplifications in the search for solutions of the equilibrium states. It is hard to decide whether one or the other are simpler or more efficient. All this aspects have to be judged depending on the particular solvable model under consideration.

Finally we note the non-solvable model situation as well. Suppose that we have a system determined by the local Hamiltonians $\{H_V(\mu)\}_V$. Then the theorem Eq. (3.14) does not hold. Nevertheless we maintain the generally valid inequality

$$f(\omega_\beta) = \inf_\omega f(\omega) \leq \inf_{\omega \in \mathfrak{Q}} f(\omega) \equiv f(\widetilde{\omega_\beta}) \tag{3.20}$$

where $\mathfrak{Q}$ is the set of homogeneous quasi-free states and $\widetilde{\omega_\beta}$ is a quasi-free state minimizing the free energy density functional for the variation over the set of qf-states. Each minimizing state can be considered as a quasi-free approximation of the true equilibrium state. Again, we proved above that the inequality becomes an equality if the system is a solvable system. If the system is not solvable then the inequality establishes a strict upper bound $f(\widetilde{\omega_\beta})$ for the free energy density $f(\omega_\beta)$ of the system. Questions can be asked about the position of the state $\widetilde{\omega_\beta}$, for example: Is it an equilibrium state and for which Hamiltonian model? The answer can be: Yes, for all systems with a solvable Hamiltonian obtained from the original one, after having applied a "quasi-free" or a "mean field" (see Eq. (4.4), Eq. (5.1)) approximation to the original model". It is well known that these quasi-free or mean field approximations are not unique or in any way canonical approximations. The system can contain many mean field approximate models of this kind.

Depending on the particular subsets of quasi-free states which we consider in these approximate variational principles, we obtain what is called in different domains of physics the usual mean field approximation models, the Hartree-Fock approximate model, the functional-density method model, the Bogoliubov model, and other solvable models; some of these models are treated in all details in Chapter Eq. (4). All these models are mean field approximations of the fully interacting two-body interaction model.

Given that the left side of the inequality Eq. (3.20) represents the real value of the free energy density of the original system, another problem appears: How closely does the right hand side value approximate the value of the left hand side, the true free energy density equilibrium value? This is again not an easy question to answer; indeed, nobody presently knows the answer. Numerous efforts to gain insight into these problems are spread throughout myriad fields of the physics literature.

Along the same line of thinking, we can consider the inf over any other subset of the homogenous states instead of solely the set of qf-states. For each of them, we again obtains upper bounds for the free energy density of the type as in Eq. (3.20). One observes that the right hand side value can greatly depend on the chosen variational set. Special sets of states which are frequently used in the literature are (sub)sets of coherent states. We recall that the special set of coherent states built on the Fock state is a subset of the quasi-free states.

We may also encounter derivations of lower bounds for the free energy density in the physics literature. It is clear that strict lower bounds can never be interpreted as the free energies of some equilibrium state of the same system. In general these lower bounds might be interesting in connection with stability questions. However in many cases it is not clear how these lower bounds present a decent physical interpretation or provide a better conceptual understanding within the variational principle scheme of statistical mechanics of boson systems.

4

Bose Einstein Condensation (BEC)

4.1 Introductory Remarks

Bose [24] and Einstein [42] considered a finite but arbitrary large set of what they called Planck oscillators and applied statistics to it. They pointed out the possibility of an arbitrary large number of oscillators to be in a zero momentum state. This property was interpreted as a physical phenomenon, a kind of condensation phenomenon. Later when second quantization was formulated, the oscillators where called boson particles. This was the birth of the famous *Bose-Einstein Condensation*, nowadays denoted in short by BEC. This work of Bose and Einstein garnered considerable discussion for more than a decade, particularly in clarifying the meaning behind the appearance of phase transitions in finite systems. In 1938 London [108] introduced the concept of macroscopic occupation of the ground state and related it to the long range coherence properties of the Bose-Einstein condensate. Since that period, the physics of the phenomenon has become standard knowledge in statistical mechanics and present in all related textbooks. Each assembly of many free boson particles shows condensation, namely a macroscopic number of the bosons in the momentum $(p = 0)$-mode, if the density is large enough or if the temperature is low enough. It is fundamentally a pure quantum phenomenon, as it holds even for a system of free quantum particles and because it disappears in the classical limit.

Apart from the free Bose gas, the interacting Bose gas comes into the picture of the BEC-phenomena as a consequence of Landau's phenomenological theory of superfluidity [95, 96]. He explained superfluidity on the basis of a quantum Bose liquid like He^4.

The so-called criterion of Landau for superfluidity is based on the idea that a quantum liquid remains a classical fluid even at zero temperature and that the classical hydrodynamical laws remain valid. The second principle is that the collective behavior dominates the movement of the individual atoms, which are the boson particles. The collective particles, sometimes called quasi-particles (see Chapter Eq. (5)), are characterized by their quasi-energies $E(k)$ with momenta $k \in \mathbb{R}^d$. *Landau's criterion for superfluidity* of the condensed matter particles at momentum $k = 0$, is obtained on a purely mechanical basis and is expressed by the condition

A.F. Verbeure, *Many-Body Boson Systems*, Theoretical and Mathematical Physics, 43
DOI 10.1007/978-0-85729-109-7_4, © Springer-Verlag London Limited 2011

$$\lim_{|k|\to 0} \frac{E(k)}{|k|} > 0 \qquad (4.1)$$

Roughly speaking, satisfying Landau's criterion requires that the spectrum of global condensate particles has a non-trivial linear behavior in the neighborhood of zero momentum. Clearly the superfluidity properties of these liquids are described in terms of the spectrum of the collective excitations.

It is important to remark at this point that the spectrum of the system enters as an important issue in the discussion about the appearance of superfluidity. Moreover, at that time, it came over as natural to take for the Bose liquid a Bose condensate. The problem with the free Bose gas as the representative model arose precisely because its spectrum did not fit with Landau's criterion for superfluidity Eq. (4.1). Therefore the search for interacting Bose gases showing Bose condensation and exhibiting the appropriate Landau spectrum came into the picture. Along this line of thought, the whole Bogoliubov theory [20, 21, 22, 169] can be seen as a tentative approach to formulating a decent microscopic theory of superfluidity. Since that time the problem concerning the occurrence of BEC for interacting boson systems has been firmly posed as a meaningful challenge to the theoretical physicists community.

The more recent interesting experiments on trapped boson gases [37, 6, 135] do not fit completely in this scheme nor in the scope of this monograph because of the presence of the external fields realizing the traps (see Section Eq. (4.8)). In these trap systems, we refer, in an intrinsic manner, to non-homogeneous systems. In this text we are interested in microscopic homogeneous boson systems. Nevertheless these trap experiments reveal appealing and interesting indications of new properties (at least, so far) of boson condensates that beg for a deeper theoretical and mathematical understanding and a modeling in terms of the standard setup of BEC for homogeneous systems. We do not treat in great detail the theory of Bose gases in traps; it remains, however, a highly active area of research. In the last section of this chapter the topic of trapped systems is shortly introduced with the following question in mind: How can these trapped-systems phenomena be understood from the microscopic point of view of homogeneous standard boson systems and their boson condensation properties?

4.2 Free Boson Gas and BEC

The free Bose gas model is described by the Hamiltonian Eq. (2.4), where the interaction potential v is set equal to zero. The local Hamiltonian defined on the Fock Hilbert space becomes

$$H_V^{free} = \int_V dx \frac{1}{2m} \nabla a^*(x).\nabla a(x) \qquad (4.2)$$

In this formulation the space derivatives enter and face us immediately with the problem of the choice of boundary conditions at the boundaries of the finite volumes V. We are also faced with the choice of the geometrical forms for these volumes. Although making choices on these points is important in the search for the possibility of BEC occurring, as a first approach we do not want to be too sophisticated in these matters. We consider periodic boundary conditions and cubic boxes. Later on we discuss in some detail the effects due to alternative choices.

Getting now to the point, consider a system of bosons of mass m enclosed in the cubic boxes $V \subset \mathbb{R}^d$, where d is the dimension of the system. The notation V denotes the subsets of the space $\mathbb{R}^d$ as well as its volume $V = L^d$; L is the side length of the boxes. We consider the dual volume

$$V^* = \left\{ k \in \mathbb{R}^d \; ; \; k_\alpha = \frac{2\pi}{L} n_\alpha \; ; \; n_\alpha = 0, \pm 1, \ldots, \; \alpha = 1, \ldots d \right\}.$$

The Hamiltonian then takes the form

$$H_V^{free}(\mu) = \sum_{k \in V^*} (\varepsilon_k - \mu) \, a_k^* a_k \tag{4.3}$$

where $\varepsilon_k = \dfrac{\hbar^2 k^2}{2m}$, $\hbar = 1$, μ is the chemical potential, and use the notation

$$a_k^* = a^*(f_k) = \int_V dx \frac{e^{ikx}}{\sqrt{V}} a^*(x)$$

$$f_k(x) = \frac{1}{\sqrt{V}} e^{ikx} \quad \text{(the individual particle wave function)}$$

with the boson commutation relations

$$[a_k, a_{k'}^*] = \delta_{k,k'} \; ; \; [a_k, a_{k'}] = 0,$$

Note that the wave functions f_k with $k \neq 0$ are periodic functions describing particular periodic localizations of the boson particles in the space variables x. On the other hand, the wave function f_0 (i.e. $k = 0$) is a constant function describing a completely delocalized one-particle wave function. This function plays a very specific role in the phenomenon of condensation; therefore many authors considered it as possibly the unique wave function of the condensed particles (at least for closed boson systems). On the other hand, particles in the other wave functions ($k \neq 0$) are usually called *excited particles*. Later we discuss situations going beyond these interpretations. For open boson systems this interpretation seems too narrow and indicates that the condensation of particles with $k \neq 0$ is also relevant in a global study of BEC.

Continuing with the free boson gas model, we first verify the stability of the free Bose gas Hamiltonian for each finite volume V. It is immediately clear that the stability criterion Eq. (3.4) for this Hamiltonian Eq. (4.3) is satisfied only when the chemical potential is strictly negative, i.e. when $\mu < 0$.

4.2.1 Standard BEC

We now look for limit Gibbs states in the finite temperature case ($\beta < \infty$) and for the appearance of condensation. We start from the finite-volume situation and look for thermodynamic-limit states of the finite-volume Gibbs states, which we denote simply by ω_V. For the Gibbs state, we then examine the particle occupation of the different energy levels for an arbitrary value of the momentum $k \in V^*$. In particular this implies already that we use periodic boundary conditions with cubic boxes of side L and finite volumes V The occupation of the number of particles is given by

$$\omega_V(N_k) = \frac{tr\, e^{-\beta H_V(\mu_V)}N_k}{tr\, e^{-\beta H_V(\mu_V)}}$$

where $N_k = a_k^* a_k$, and easily computed to be

$$\omega_V(N_k) = \frac{1}{e^{\beta(\varepsilon_k - \mu_V)} - 1}.$$

The chemical potential μ_V of the volume V is determined by the fixed given total density constraint (grand canonical ensemble) equation for each volume V,

$$\rho = \omega_V(N_V) = \frac{1}{V} \sum_{k \in V^*} \frac{1}{e^{\beta(\varepsilon_k - \mu_V)} - 1} \tag{4.4}$$

Note that this relation depends on the size and form of the volume and therefore everything is boundary conditions dependent. In order to make more universal statements (more precisely, to make things less dependant on the size of the volume) we take the *thermodynamic limit*. We let L or the volume tend to infinity while keeping the density $\rho = \omega_V(N_V)$ constant. Then for each volume V we fix a value of the chemical potential $\mu_V(\beta, \rho)$ and vice versa. Let us consider now the thermodynamic limit of this density equation, using the notation $\mu \equiv \mu(\beta, \rho) = \lim_{V \to \infty} \mu_V \leq 0$. We then obtain

$$\rho = \lim_{L \to \infty} \omega_V\left(\frac{\langle N_V \rangle}{V}\right) = \lim_{L \to \infty} \left(\frac{1}{V} \frac{1}{e^{-\beta \mu_V} - 1} + \frac{1}{V} \sum_{k \neq 0} \frac{1}{e^{\beta(\varepsilon_k - \mu_V)} - 1} \right)$$

$$\rho = \rho_0(\beta, \rho) + \rho(\beta, \mu(\beta, \rho)) \tag{4.5}$$

with

$$\rho(\beta, \mu) = \left(\frac{1}{2\pi}\right)^d \int dk \frac{1}{e^{\beta\left(\frac{k^2}{2m} - \mu\right)} - 1}$$

$$\rho_0(\beta, \rho) = \lim_{L \to \infty} \frac{1}{V} \frac{1}{e^{-\beta \mu_V} - 1} = \lim_{L \to \infty} \omega_V\left(\frac{a_0^* a_0}{V}\right).$$

Clearly $\rho_0(\beta, \rho)$ is the density of particles in the one-particle wave function vector of lowest energy $\varepsilon_0 = 0$ for a fixed total density ρ.

As the function $\mu \to \rho(\beta, \mu)$ increases monotonically we obtain

$$\rho(\beta,\mu) \leq \rho(\beta,0) \equiv \rho_c(\beta) = \left(\frac{1}{2\pi}\right)^d \int dk \frac{1}{e^{\beta \frac{k^2}{2m}} - 1}. \tag{4.6}$$

One checks that $\rho_c(\beta)$, called the *critical density* of the free boson gas, is a finite function of β for all dimensions $d > 2$ and that it diverges for dimensions $d = 1, 2$. As we will see later, dimensionality turns out to be important for the occurrence of BEC. Clearly the story stops here in the sense that there is no condensation except when we consider systems with dimensions $d \geq 3$. For convenience, we should keep in mind the case of dimension $d = 3$ or $d \geq 3$ to obtain a finite critical density.

Suppose first that the fixed total density ρ is small. In particular suppose that the density is smaller than the critical density, $\rho < \rho_c(\beta)$; then there exists a unique chemical potential $\mu \equiv \mu(\beta,\rho) < 0$ such that $\rho = \rho(\beta,\mu(\beta,\rho))$ and that the total density equation Eq. (4.4) is satisfied with $\rho_0(\beta,\rho < \rho_c) = 0$.

On the other hand, if the total density ρ is large enough, in particular if it is larger then the critical density, $\rho > \rho_c(\beta)$, the density constraint equation still holds in the form

$$\rho - \rho_c(\beta) = \rho_0(\beta,\rho) > 0 \tag{4.7}$$

The chemical potential vanishes, that is $\mu = 0$, and we obtain a macroscopic occupation of particles in the ground ($k = 0$) energy level. If so, Bose-Einstein condensation has taken place in the ($k = 0$) mode.

Before going on, let us make a technical remark about the limit two-point function $\omega(a^*(f)a(g))$, for any couple f,g of the test function space $\mathscr{S}$. Let the state ω be the equilibrium state or (thermodynamic) limit Gibbs state for the free Bose gas. We obtain the following two-point function result as a consequence of the possible condensation phenomenon, which we based on the fixed total density equation (4.7) argument. Indeed this equation yields straightforwardly that the two-point function is of the form

$$\omega(a^*(f)a(g)) = \widehat{f}(0)\overline{\widehat{g}}(0)\rho_0 + \left(\frac{1}{2\pi}\right)^d \int dk \, \widehat{f}(k)\overline{\widehat{g}}(k) \frac{1}{e^{\beta \frac{k^2}{2m}} - 1}$$

This means that the two-point function of the free boson gas is continuous with respect to the following norm on the test function space $\mathscr{S}$:

$$\|f\|_0^2 = |\widehat{f}(0)|^2 + \int dk \, |\widehat{f}|^2(k) \frac{1}{e^{\beta \frac{k^2}{2m}} - 1} \tag{4.8}$$

Therefore the two-point function extends to the closure $\overline{\mathscr{S}}$ of $\mathscr{S}$ with respect to this new norm, showing the particular status of the zero or condensate mode. The zero mode becomes a macroscopic mode with a status different from all other k-modes. We can formulate this property as follows: In the thermodynamic limit the zero-mode becomes an independent macroscopic supplementary degree of freedom separated from the other degrees of freedom of the system. This fact becomes even more explicit in the next section when we consider the condensate equation valid for general interacting boson systems.

Let us analyze further the situation of condensation. Suppose that we have indeed a non-trivial condensate $\rho_0 > 0$, or condensation in the mode $k = 0$. Then from the total density equation we get that

$$0 < \rho_0 = \lim_{L \to \infty} \frac{1}{V} \frac{1}{e^{-\beta \mu_V} - 1} = \lim_{L \to \infty} \frac{1}{V} \left(\frac{1}{-\beta \mu_V} \right) \leq \rho$$

or equivalently we find that the chemical potential μ_V behaves at large volumes as $\mu_V \simeq -1/V$. We derived the rate of convergence to zero of the chemical potential as a function of the volume ultimately leading to $\lim_V \mu_V = 0$. We determined the large volume dependence of the chemical potential.

Our next remark concerns the question of condensation outside of the ground state mode. In fact we can easily understand why there should be no condensation or macroscopic occupation in any of the other modes $k \neq 0$ for the free Bose gas with periodic boundary conditions. Let us look at what happens with the occupation in the first excited mode given by $|k| = \frac{2\pi}{L}$. The analysis for higher modes follows along the same lines. Looking at the one-particle energies of the excited modes we find for $d = 3$,

$$\inf_{k \neq 0} \frac{k^2}{2m} = \frac{1}{2m} \left(\frac{2\pi}{L} \right)^2 \inf_{n \neq 0} \left(n_1^2 + n_2^2 + n_3^2 \right) \simeq \frac{1}{V^{2/3}} \; ; \; \varepsilon_k \simeq \frac{1}{V^{2/3}}$$

and therefore the particle occupation of this lowest non-zero mode in the limit $L \to \infty$ tends to zero. Indeed

$$\lim_{L \to \infty} \frac{1}{V} \frac{1}{e^{\beta \varepsilon_k} - 1} \bigg|_{k \neq 0} \simeq \lim_{L \to \infty} \frac{1}{V} \frac{1}{\left(\frac{1}{V^{2/3}} \right)} = \lim_{L \to \infty} \frac{1}{V^{1/3}} = 0.$$

In conclusion, there can be macroscopic occupation in the $(k = 0)$-mode, but no macroscopic occupation for any of the other modes $k \neq 0$.

In conclusion we proved the existence of BEC for the free Bose gas in the lowest energy $(k = 0)$-mode and the absence of condensation in the $(k \neq 0)$-excited modes. These phenomena show up when the particle density is high enough, a condition which can also be replaced by the condition of sufficiently low temperature, because the critical density decreases monotonically as a function of inverse temperature. This result holds for all dimensions $d \geq 3$. In the proofs we used essentially the so-called saturation argument, expressing that all particles beyond those needed to reach the critical density should condense. In this analysis we used periodic boundary conditions, cubic boxes, and a thermodynamic limit such that the side L of the box tends to infinity. We also considered the thermodynamic limit of the finite volume Gibbs states.

Our arguments do not work for dimensions $d = 1, 2$. In that case it follows from our argument that there is no condensation. At this point it is good to mention the huge literature dealing with disproving condensation, in particular dimensions $d = 1, 2$, even for boson systems with interactions. All argumentations in this direction are based on the Mermin-Wagner argument [120, 167], which is a generalization of the saturation argument used above, but which is expressed in a form applicable to interacting systems as well. It is one of the commonly known exact results in the many-body boson theory. In Eq. (4.3.1) we discuss this argument in a more detailed form.

Before proceeding to the next topic, we note that Landau's criterion for super-fluidity Eq. (4.1) is not satisfied for the free boson gas with condensation. The free boson condensate particles coincide with the original particles in the sense that they have the same spectral energy function. Indeed one has $E(k) = \varepsilon_k = \frac{k^2}{2m}$ and therefore $\lim_{|k| \to 0} \frac{\varepsilon_k}{|k|} = 0$, contradicting Landau's criterion.

4.2.2 Thermodynamic Limit and Boundary Conditions

In this section we want to draw attention to the fact that the arguments leading to the results of our analysis above do not provide the only possible description of the occurrence of Bose-Einstein condensation. The contributions from the research in mathematical physics, in particular by the Dublin group (see e.g. [101, 18, 17]) are important in this respect. They illustrate the dependence on the type of the boundary conditions and on the type of thermodynamic limits at several places in the analysis of the free boson gas.

Consider first the dependence of this description on the type of thermodynamic limit. In this context the concept of generalized condensation is discovered and worked out by several authors. *Generalized condensation* is condensation appearing in modes $k = \frac{2\pi n}{L} \neq 0$, but which of course have the property of tending to zero along with the thermodynamic limit. This type of condensation can occur accompanied or not with condensation in the mode $k = 0$. For a literature review of this phenomenon, see [169]. We should stress that this type of condensation is not just a mathematical artifact; it is also experimentally observed and it is called, in experimental physics, the phenomenon of *fragmentation*. Without going into all details about this subtle phenomenon [16, 14] we illustrate it on the basis of taking an example of the thermodynamic limit where we use special types of volumes not to much different from simple cubic boxes. Let us consider, for example, increasing rectangular boxes of the type $V = L_1 \times L_2 \times L_3$ and we work in dimension $d = 3$. The volumes are selected according to the following types:

$$V = V^{\gamma_1} V^{\gamma_2} V^{\gamma_3} \text{ with } \gamma_1 + \gamma_2 + \gamma_3 = 1 \text{ and } V^{\gamma_i} = L_i$$

Considering cubic boxes means that we take $\gamma_1 = \gamma_2 = \gamma_3 = 1/3$. But now let $\gamma_1 = 1/2$ and $\gamma_2 = \gamma_3 = 1/4$. With the last choice the one-particles energies become

$$\frac{k^2}{2m} = \frac{1}{2m}(2\pi)^2 \left(\frac{n_1^2}{V} + \frac{n_2^2}{V^{1/2}} + \frac{n_3^2}{V^{1/2}} \right) \; ; \; n_\alpha = 0, \pm 1, \dots .$$

The density constraint relation Eq. (4.4) in the finite volumes V produces the explicit expression

$$\rho = \frac{1}{V} \sum_{\substack{n_1 = 0, \pm 1, \dots \\ n_2 = n_3 = 0}}^{\prime} \frac{1}{e^{\beta(\varepsilon_k - \mu_V)} - 1} + \frac{1}{V} \sum_{\substack{n_\alpha, n_{\alpha \neq 1} \neq 0}}^{\prime\prime} \frac{1}{e^{\beta(\varepsilon_k - \mu_V)} - 1} .$$

Let us again consider the limit $L \to \infty$. Because μ_V is still of the order $-\frac{B}{V}$ in the volume V, for $\rho > \rho_c(\beta)$, we find

$$\rho - \rho_c(\beta) = \lim_L \frac{1}{V} \sum' \frac{1}{e^{\beta \varepsilon_k} - 1} = \sum_{n_1 = 0, \pm 1, \ldots} \frac{1}{A n_1^2 + B} = \rho_0 > 0$$

with A and B positive constants. This expression proves that there is condensation in infinitely many modes. Clearly we proved this effect on the basis of considering a particular type of thermodynamic limit. This simple example illustrates that condensation solely in the $(k = 0)$-mode is a much more subtle affair than what may be expected from the analysis with periodic cubic boundary conditions. Moreover, in [123] a model with periodic boundary conditions is considered showing BEC in infinitely many modes, but without condensation in the mode $k = 0$. In this model the addition of a simple repulsive interaction term to the mean field Bose gas Eq. (4.4) is responsible for this phenomenon of generalized condensation. In other words, the authors show that generalized condensation can also show up as a consequence of the presence of special types of interactions.

Now we discuss the influence of different boundary conditions and consider therefore again the free Bose gas, not with periodic boundary conditions as we did above, but with what is called *attractive boundary conditions* ([147, 160]). Consider the free particle moving in the one-dimensional interval [-L/2,L/2] (system dimension $d = 1$). We look first for the spectrum of the Laplace operator and consider its spectrum, which is now the appropriate one-particle spectrum of the model.

$$-\frac{d^2 \phi(x)}{dx^2} = \lambda \phi(x)$$

The attractive boundary conditions are defined by the equations

$$\left(\frac{d\phi}{dx} + \sigma\phi\right)\Big|_{x=-\frac{L}{2}} = 0; \; \left(\frac{d\phi}{dx} - \sigma\phi\right)\Big|_{x=\frac{L}{2}} = 0$$

The positive number σ is called the *elasticity parameter* of the boundaries. Since we are interested in the thermodynamic limit we can limit ourselves to the intrinsic case $L\sigma > 2$, and in that case we obtain two strictly negative eigenvalues $\varepsilon_0(L) \le \varepsilon_1(L) < 0$, both tending to $-\sigma^2$ for L tending to infinity. The whole one-particle spectrum can be ordered as follows

$$\varepsilon_0(L) \le \varepsilon_1(L) \le \varepsilon_2(L) \le \varepsilon_3(L) \le \ldots$$

We can check that the volume or the L-dependence of the eigenvalues is given by $\varepsilon_0(L) = -\sigma^2 - O(e^{-L\sigma})$ and $\varepsilon_1(L) = -\sigma^2 + O(e^{-L\sigma})$ for the negative eigenvalues; for the positive eigenvalues $k \ge 2$, we have $((k-1)\pi/L)^2 \le \varepsilon_k(L) \le (k\pi/L)^2$.

We note the main difference with respect to the situation of periodic boundary conditions. In the attractive boundary conditions case the one-particle spectrum shows a finite gap between the two negative energies, which coincide in the limit $L \to \infty$, and the non-negative energies.

The corresponding eigenfunctions $(\phi_k)_k$ can also be computed explicitly and are, for $k = 0, 1$, given by

$$\phi_0^L(x) = \sqrt{\frac{2}{L}}(1 + \frac{\sinh L\sigma}{L\sigma})^{-1/2}\cosh(-\sigma x)$$

$$\phi_1^L(x) = \sqrt{\frac{2}{L}}(-1 + \frac{\sinh L\sigma}{L\sigma})^{-1/2}\sinh(-\sigma x)$$

For even $k \geq 2$

$$\phi_k^L(x) = \sqrt{\frac{2}{L}}(1 + \frac{\sin\sqrt{\varepsilon_k(L)}L}{\sqrt{\varepsilon_k(L)}L})^{-1/2}\cos(\sqrt{\varepsilon_k(L)}x)$$

and for odd $k \geq 3$

$$\phi_k^L(x) = \sqrt{\frac{2}{L}}(1 - \frac{\sin\sqrt{\varepsilon_k(L)}L}{\sqrt{\varepsilon_k(L)}L})^{-1/2}\sin(\sqrt{\varepsilon_k(L)L}x)$$

The eigenfunctions for the negative energies are hyperbolic functions yielding a maximal density of particle near the boundaries. For positive energies we find the usual harmonic functions, as in the periodic boundary conditions situation. To look for the occurrence of BEC in this model it is sufficient to repeat the saturation argument developed above in the case of periodic boundary conditions. Using the explicit form of the spectrum and the eigenfunctions, and after performing the limit $L \to \infty$, we can readily compute a critical density , which is given by the following expression containing explicitly the elasticity parameter σ of the walls:

$$\rho_c(\beta) = \frac{2}{\pi}\int dk \frac{1}{e^{\beta(\frac{k^2}{2m} + \sigma^2)} - 1}$$

Because $\sigma^2 > 0$ this integral remains finite in all dimensions. Hence for total densities ρ larger than this critical density we get condensation, consisting of macroscopic occupation of the lowest energy level with a condensate density which again is given by the formula: $\rho_0 = \rho - \rho_c(\beta)$. Clearly this model illustrates explicitly that the occurrence of BEC indeed depends greatly on the choice of boundary conditions. All this should warn us to take care of the boundary conditions in general and in particular when we refer to so-called basic theorems in physics, like the Mermin-Wagner one [120] which we will discuss later. Otherwise there are also many results dealing with proofs about the absence of condensation. It is typical that many of them include the explicit specification of the boundary conditions, see e.g. [48]. In this context we should mention an interesting paper [76] which uses the free boson gas model with attractive boundary conditions as a model to search for equilibrium states containing the possibility of a microscopic explanation of the condensate vortices. The ideas developed in this paper could contain the germ for interesting future exploitations about this topic, which should hold in the case of interacting boson systems as well.

Another type of boundary condition studied in some detail in the literature is so-called *scaled weak external fields* [15, 136]. The trapped boson models Eq. (4.8) are also external field models, but in these models the field is not scaled and therefore these systems are inhomogeneous. The scaled external field boson models have a one-particle Hamiltonian of the following type in, for example, one spacial dimension:

$$h_L = -\frac{1}{2m}\frac{d^2}{dx^2} + v(\frac{x}{L}) - \mu \tag{4.9}$$

The external potential v is a non-negative continuous function with a global minimum at $x = 0$, where $v(0) = 0$. The stability of the system requires again that the chemical potential at each finite volume is strictly negative $\mu_V < 0$. Following again the saturation argument, we obtain a critical density now given by a different expression

$$\rho_c(\beta) = \frac{1}{2}\int\frac{dk}{2\pi}\int_{-1}^{1}dq\{\exp\beta(\frac{k^2}{2m} + v(q)) - 1\}^{-1}$$

which depends on the external field. It is clear that this critical density is finite even in one dimension for many non-trivial external potentials v and that the existence of condensation occurs again for sufficiently high total particle densities. For more details on these type of models, see Section Eq. (4.8).

All these models indicate the possibility for the occurrence of BEC of highly distinct natures depending greatly on the type of boundary conditions and on the specific thermodynamic limits which are considered.

To partly avoid some of these subtle effects it is perfectly reasonable to work immediately in the thermodynamic limit with properly adapted definitions of equilibrium states. That is precisely the reason why we introduced in Chapter Eq. (3) the variational principle of statistical mechanics and the energy-entropy balance conditions formulated in the thermodynamic limit to define equilibrium states.

Finally, it is clear that the free Bose gas in a thermodynamic limit of any type of boundary condition is a solvable model. Therefore, as well the variational principle on the set of quasi-free states Eq. (3.14), as the energy-entropy balance criterion for equilibrium states Eq. (3.7), should allow for an explicit and complete computation of all the equilibrium states. In particular both of these criteria should lead to the proof of the existence of BEC for large particle densities. We leave the details of these derivations or proofs as an exercise. (The solution of this exercise in Section Eq. (4.4) provides a hint.)

Before we start the detailed study of a number of other solvable boson models for which we exploit these equilibrium criteria, we proceed in the section to the derivation and discussion of a number of universal exact results holding for arbitrary and fully interacting two-body boson systems. In other words we derive and discuss universal basic properties about the equilibrium states displaying or not displaying Bose-Einstein condensation.

4.3 BEC in Interacting Boson Gases

When we refer to an interacting Bose gas we have in mind a system defined by a super-stable Hamiltonian Eq. (2.4) with a non-trivial two-body potential v. In general this is a non-solvable problem in the sense that up to now nobody has been able to derive from the variational principle Eq. (3.1), or from the energy-entropy balance criterion Eq. (3.7) for equilibrium, explicitly one or all ground states and/or equilibrium states. Even in this situation, it does not mean that nothing can be said about these sofar not explicitly known states. To the contrary, any result directly and rigorously derived from the equilibrium criteria is of universal importance and useful for the understanding of real boson systems behavior. Indeed researchers have been able to derive from the criteria a number of exact results about possible equilibrium or ground states. In this section we give a discussion of the basic Mermin-Wagner argument, proving in general the absence of condensation in some situations. We focus also on the intrinsic basic relation between the occurrence of boson condensation and the occurrence of spontaneous symmetry breaking and off-diagonal long range order. Because of the importance of this property reaching far beyond boson systems, the latter subject is extensively worked out. Finally we derive rigorously from the equilibrium criteria the so-called condensate equations. These are independent equations for one or more condensate densities or order parameters. All these general properties are illustrated for a number of relevant interacting boson theories or boson models in the next subsections.

4.3.1 Mermin-Wagner Argument

In the literature one of the most widespread cited theorems about boson condensation sounds as follows: In one and two dimensions the gauge symmetry of any boson system model cannot be spontaneously broken (see next subsection). Or expressed otherwise, in these low dimensions no condensation can occur at any finite positive temperature. It sounds as: Boson condensation is a phenomenon of dimensions three or higher. The reason for this adagio is often expressed in words as follows: if such a symmetry breaking would occur, the emerging (quasi-)particles would have a non-integrable correlation function. Sometimes this is also argued by saying that in one and two dimensions the temperature fluctuations would destroy the condensate.

The original Mermin-Wagner argument [120] was first successfully applied in the Heisenberg quantum spin model. However, it was clear that the argument was of a much more general validity. Hohenberg [77] applied it to Bose and Fermi systems and showed that there is no boson condensation, respectively superconductivity possible in dimensions $d = 1$ or $d = 2$. This result is always referred to as an exact result because it is based on an exact inequality, originally due to Bogoliubov (see Eq. (3.9) and Eq. (3.11)).

Nevertheless it must be remarked that the generality of some aspects of this result has been questioned in the literature at many occasions. Already within the scope of this text on Bose systems it was pointed out, as a rigorous result, that Bose condensation can occur in one and two dimensions, for instance in systems with attractive boundary conditions [147] or in systems with weak external fields [15, 136]. It is good to keep in mind that these results hold already for the free boson gas. Therefore, giving without care the powerful and valuable argument of Mermin-Wagner, the connotation of a theorem is a bit hazardous. Maybe each theorem should always be accompanied by a discussion about the conditions under which the statement really holds. In particular for bosons, but even for spin systems, the boundary conditions do play an important role in the argumentations. However for periodic boundary conditions (for physicists the common sensical choice for boundary conditions) the Hohenberg result holds true. For completeness we reproduce it here as the following theorem:

Theorem 4.1. *Consider an interacting Bose gas (Eq. (2.4) and Eq. (2.6)) with a potential term satisfying the stability condition, with periodic boundary conditions and with a thermodynamic limit with boxes. Let ω be any ergodic equilibrium state with a finite particle density ρ. Then for all positive temperatures there is absence of condensation in one and two dimensions.*

Proof. Remember first that for ergodic states $\rho_0 = \lim_V |\omega(a_0/\sqrt{V})|^2$. Therefore the absence of condensation ($\rho_0 = 0$) is equivalent to the absence of the gauge breaking in the one-point function, that is, with $\lim_V \omega(a_0/\sqrt{V}) = 0$.

The essential ingredient for the proof of this theorem is the inequality of Bogoliubov Eq. (3.11), which holds for any equilibrium state ω and is written in the form

$$|\omega([A,B])|^2 \leq \frac{\beta}{2}\omega([A^*,[H_V(\mu),A]])\omega(B^*B+BB^*)$$

for any pair of observables A and B, where again $H_V(\mu) = H_V - \mu N_V$.

Consider any finite volume V, periodic boundary conditions, and any value of the momentum $k \neq 0$. For A and B let $A = \sum_{k'} a_{k'}^* a_{k'-k}$ and $B = a_k$. We can calculate straightforwardly the following commutators: Denote for short by U_V the potential term of the Hamiltonian and by T_V the kinetic energy, and compute the commutators $[A,B] = -a_0$ and $[A,U_V] = [A,N_V] = 0$. Therefore

$$[A^*,[H_V(\mu),A]] = \frac{k^2}{m}N_V$$

After substitution in the Bogoliubov inequality we obtain

$$\frac{|\omega(a_0/\sqrt{V})|^2}{k^2} \leq \frac{\beta}{2m}\omega(a_k^*a_k + a_k a_k^*)\rho$$

Consider now any $\varepsilon > 0$. Let $S_1(V)$ be the unit sphere of the dual V^* of the volume V and $S_1(V)_\varepsilon$ the unit sphere excluding the sphere of radius ε and use the resulting inequality to obtain

$$\frac{1}{S_1(V)_\varepsilon}\sum_{k \in S_1(V)_\varepsilon} \frac{|\omega(a_0/\sqrt{V})|^2}{k^2} \leq \frac{\beta\rho}{2mS_1(V)_\varepsilon}\sum_{k \in S_1(V)_\varepsilon}\omega(a_k^*a_k + a_k a_k^*)$$

Consider the infinite volume limit of this inequality. With the notation $\lim_V S_1(V)_\varepsilon = S_{1,\varepsilon}$, we obtain

$$\lim_V \int_{S_{1,\varepsilon}} dk \frac{|\omega(a_0/\sqrt{V})|^2}{k^2} \leq \frac{\beta\rho}{2m}\int_{S_{1,\varepsilon}} dk\, \omega(a_k^*a_k + a_k a_k^*)$$

The right hand side of this inequality is, for all values of ε, bounded by a constant $C(\rho)$ that depends only on the density ρ. Using again the ergodicity of the state such that $\lim_V |\omega(a_0)/\sqrt{V}|^2 = \rho_0$, we now obtain

$$\rho_0 \int_{S_{1,\varepsilon}} dk \frac{1}{k^2} \leq \frac{\beta \rho C(\rho)}{2m}$$

The upper bound is finite for all finite densities ρ. The lower bound integral however is divergent in one or two dimensions for ε tending to zero. Hence, if $\rho_0 > 0$ then the exact inequality is violated. Therefore the only possibility to get out of this deadlock is for $\rho_0 = 0$, which means the absence of condensation and/or the absence of spontaneous gauge symmetry breaking. This finishes the proof of the theorem.

Of course the Mermin-Wagner argument holds true as a theorem and is proved for considerably more types of boundary conditions than just the periodic ones. The literature contains several theorems which differ only in the type of boundary conditions assumed. Comparing all these results is not a straightforward business because in general the boundary conditions problem for quantum systems is a complex technical subject and even as an independent subject not fully clarified (see [56] for example). Moreover many of the existing results are model dependent. Nevertheless some of us might be interested in getting a precise idea where and how the types of boundary conditions precisely enter in the proofs of the theorem. We can find more information when one looks at the details of the paper [48].

4.3.2 Spontaneous Symmetry Breaking (SSB) and BEC

Again we consider fully interacting boson systems, which are defined by the local Hamiltonians H_V Eq. (2.4) where the V are the finite volumes of $\mathbb{R}^d$. These Hamiltonians share the property of space translation invariance and satisfy the property $\tau_x H_V = H_{V+x}$, where $V + x$ is the set V translated over the distance $x \in \mathbb{R}^d, d \geq 1$. We say that the systems have *space translation symmetry*, the symmetry group is the group $\mathbb{R}^d$ of translations.

Our systems under consideration are also gauge invariant Eq. (2.28). This is expressed by $\tau_\lambda H_V = H_V$ for all $\lambda \in [0, 2\pi]$ because the Hamiltonian contains only sums of terms which are products of an equal number of creation and annihilation operators. Our system has the *gauge transformation symmetry*. The symmetry group is in this case the simplest non-trival unitary group $U(1)$.

It follows immediately that all Gibbs states ω_V Eq. (3.5) share as well the space translations as the gauge transformation symmetries. Also their thermodynamic limit Gibbs states $\omega = \lim_{V \to \infty} \omega_V$ share these symmetries and satisfy the equilibrium conditions Eq. (3.1), as well as Eq. (3.7).

In fact we are interested in all possible limit equilibrium states, that is, all solutions of at least one of the two equilibrium conditions. In this subsection we choose for the energy-entropy balance conditions Eq. (3.7) as our equilibrium criterion. We leave the derivation of these same results as explained below, but starting from solutions of the variational principle Eq. (3.1), as an exercise.

It turns out that under the special circumstances of low temperature and/or high total density, we can find states which are solutions of the energy-entropy balance criterion of equilibrium Eq. (3.7) which are or are not space translation symmetry invariant, but which are in any case not gauge transformation invariant, although the local Hamiltonians as well as their limit Gibbs states do always have these symmetries. If this happens, we speak of *spontaneous symmetry breaking*(SSB).

We should note that SSB can occur only in the thermodynamic limit formulation of equilibrium, as a clear consequence of Theorem Eq. (3.5). It is also clear that adding a symmetry breaking term to the Hamiltonian is a forced way of breaking the symmetry of the system and is not the same as the phenomenon of spontaneous breaking of the symmetry.

Because of the great impact of the SSB-subject on modern physics in general and not only on many-body boson systems, we describe first the problem in a more explicit and in a more sophisticated way which could be appreciated also by the more mathematically-schooled expert interested in the phenomenon of SSB. In other words, the underlying sufficient mathematical conditions will be rather extensively and explicitly indicated. Afterwards it is made clear that there are no real physical conditions to be formulated, except to start with a finite volume equilibrium Gibbs state whose thermodynamic limits ω_β exist. We are interested in all states satisfying the energy-entropy balance conditions Eq. (3.7) and which are related to the limit Gibbs states ω_β. As explained before, the latter are also solutions of Eq. (3.7) and are translation and gauge invariant. Therefore it is reasonable to start with any solution ω of Eq. (3.7), independent of the fact that it is a limit Gibbs state or not, which satisfies the following properties automatically satisfied by thermodynamic limit Gibbs states: The initial situation is therefore that each initial state ω considered satisfies the properties:

A. The state ω is translation and gauge invariant
B. The state ω is analytic in the sense of [26] p.38. This is a technical term used simply to express the fact that the state can be completely described in terms of all (n, m)-point correlation functions of the creation and annihilation operators.

In fact all of our theorems remain true even without imposing the second condition **B**. In that case we should consider only the observables written in terms of the Weyl operators and rewrite the equilibrium conditions in a technically somewhat more sophisticated, but physically less transparent, form. However that is something which we prefer not to do.

In this subsection we describe the solution of the main problem: Suppose we are given such a (limit Gibbs) ω state ω. Can we always find or construct its ergodic components equilibrium states? Do all the latter states have the same symmetries as the given (limit Gibbs) state or does spontaneous symmetry breaking occur, and under which circumstances does one and the other happen? We provide and explain the full solutions needed to answer these questions. In short, the answers contain an explicit construction of ergodic equilibrium states that break the gauge symmetry, and in some cases also the translation symmetry, if and only if there is boson condensation. An explicit expression for these symmetry breaking states is written down in terms of the initially given (limit Gibbs) state ω.

Let us go into some more details. We consider again periodic boundary conditions for the finite volume argumentations. The state ω shows *condensation* in the mode $q \in V^*$ if

$$\rho_q = \lim_V \omega \left(\frac{a_q^* a_q}{V} \right) > 0 \qquad (4.10)$$

with ρ_q called the *q-condensate density*.

For homogenous closed boson systems the condensation in $q = 0$, called *ground state condensation*, is the best known and intensively studied. See for example the treatment of the free Bose gas. Looking at the literature about boson condensation of these homogenous systems, we would notice on many occasions that numerous authors seem to take for granted, or consider it as a trivial fact, that the existence of ground state condensation (i.e., $\rho_0 > 0$) implies spontaneous gauge symmetry breaking and vice versa. Sometimes they refer to this situation by writing down blindly

$$\rho_0 = \lim_V \omega \left(\frac{a_0^* a_0}{V} \right) = \lim_V \left| \omega \left(\frac{a_0}{\sqrt{V}} \right) \right|^2 \qquad (4.11)$$

mostly coming out of the blue. Probably the reason for this conclusion is that the statement, breaking of the gauge symmetry implies condensation, is an immediate property. Indeed this statement follows immediately from the Schwartz inequality which holds for any state and yields: $|\omega(a_0)|^2 \leq \omega(a_0^* a_0)$. However the non-trivial point is the inverse inequality which does not hold in general for all states.

The point with Eq. (4.11) is that for a limit Gibbs state ω being gauge invariant and therefore yielding $\lim_V \omega(a_0/\sqrt{V}) = 0$, we get immediately a fatal contradiction if there is condensation or if $\rho_0 > 0$. On the other hand the equality Eq. (4.11) is obviously valid for all ergodic states. In particular the equality Eq. (4.11) expresses precisely the ergodicity of the state for the 0-mode subsystem. The point is that limit Gibbs states need not be ergodic in all circumstances. To the contrary, if there is condensation this is particularly not the case for all values of the total density or for all values of the temperature. Indeed if a phase transition is showing up, like condensation coming around the corner, we cannot expect that the limit Gibbs states are ergodic for the condensate mode. The state might be a mixture of a condensate and a non-condensate phase state. The aim of this subsection is precisely to explain how this puzzle is solved.

This problem was solved rigorously already some time ago in a very general setting in [49], where only the case of ground state condensation was considered, resulting in the following theorem: If there is a limit Gibbs state ω showing condensation, $\rho_0 > 0$, then there exists always a family of gauge breaking equilibrium states $\{\omega_\alpha(\cdot)\}_{\alpha \in [0,2\pi]}$ such that the following two properties are satisfied:

(a) $\omega(\cdot) = \frac{1}{2\pi} \int_0^{2\pi} d\alpha \, \omega_\alpha(\cdot)$

(b) $\lim_V \left| \omega_\alpha \left(\frac{a_0}{\sqrt{V}} \right) \right| = \lim_V \omega_\alpha \left(\left(\frac{a_0^* a_0}{V} \right)^{\frac{1}{2}} \right) \neq 0$ for all values of α

The first equality tells us that the limit Gibbs state is a non-trivial convex combination of the states ω_α and therefore is not ergodic. The second one expresses the ergodicity of all the states ω_α. This property follows as a consequence of the explicit construction of the states ω_α given in terms of the given state ω. We reproduce the proof of this result, which consists essentially in the explicit construction of the gauge breaking equilibrium states $\{\omega_\alpha\}$ out of our given limit Gibbs state ω in the case that the latter one is non-ergodic. We show that the last property follows if the system shows Bose-Einstein condensation. It is clear that this construction is the key point contribution to the relation between spontaneous symmetry breaking and condensation. How do the ω_α look like? Impatient ones among us can glance at Eq. (4.22) below, where they can see the explicit construction of these gauge breaking equilibrium states in terms of the given limit Gibbs state and the condensate density. These states are explicitly given in terms of their (n,m)-point correlation functions. The simplicity of the formula Eq. (4.22) defining these states might motivate them to undergo the full proof dealing with its derivation and its correctness. In what follows we will provide in full detail the whole mathematical argumentation to arrive at this formula as well as for its generalization to the case of non-ground state condensation.

It might look surprising that we include in this text the case of this more general q-condensation. This surprise may come up in view of the elegant result in [153] where a proof regarding the absence of currents is obtained for the BCS-model. Our argument for the inclusion of this generalization is motivated by the recent activity in the domain of condensation in traps [6, 37], where this type of non-ground state condensation ($q \neq 0$) is relevant. These kind of systems have indeed to be classified among the open boson systems driven by another external system. Therefore we reproduce also the generalized theorem, first proved in [138]. It makes clear that, in the case of $q \neq 0$, not only the gauge symmetry is spontaneously broken, but also the space translation symmetry. The main new result is that for homogenous gauge invariant systems, the thermodynamic limit Gibbs states, showing condensation in a mode $q \neq 0$, can be decomposed with respect to periodic equilibrium states showing also gauge symmetry breaking. We get in this case SSB of the space and gauge symmetries. The obtained decomposition of the state ω in one q- space-like direction, fixed by the unit vector $\hat{\mathbf{e}} = \mathbf{q}/|q|$, can straightforwardly be generalized to more or all dimensions by constructing again new equilibrium states. The latter ones are periodic in more, respectively all dimensions and they are gauge breaking in all these q-modes which fix the different directions. So far for the setting and the essentials of the content of what follows below.

Technicalities

We start the construction with the one-momentum vector case, $q = (2\pi/\gamma)\hat{\mathbf{e}}$, where $\hat{\mathbf{e}}$ is a unit vector of $\mathbb{R}^d$ and where γ is the *period or wavelength* determined by the vector q. In what follows ω will be an equilibrium state, satisfying the equilibrium conditions Eq. (3.7) and the conditions A (gauge and translation invariant) and B above. The main result yields that, if such a state ω with q-condensation exists, then there exist gauge-breaking periodic states with finite period γ if $q \neq 0$. If $q = 0$, the states are gauge breaking but remain homogeneous.

For convenience we repeat the mathematical technology offered by the GNS-construction (see Eq. (7.1)). We consider the boson system $(\mathfrak{A}, \omega)$ and its GNS-representation, which is an algebra representation into the linear operators on the Hilbert space $\mathscr{H}$ which contains a cyclic and separating vector $\Omega \in \mathscr{H}$, such that, for all $X \in \mathfrak{A}$ holds

$$\omega(X) = (\Omega, X\Omega)$$

In this formula we used for notational convenience the same symbol for the element X of the algebra $\mathfrak{A}$ and for its image under the GNS-representation which is an operator acting on the Hilbert space $\mathscr{H}$. The cyclicity property of the vector Ω means that the set of vectors $\{X\Omega | X \in \mathfrak{A}\}$ is dense in $\mathscr{H}$. Meanwhile we note once again that the GNS-representation theorem proves that each state can indeed be interpreted as an expectation-valued map as explained in Chapter Eq. (2) as well as in the Appendix Eq. (7.1).

Consider the given vector $q \in V^*$ and the operator $\alpha_{q,V}$ acting on $\mathscr{H}$

$$\alpha_{q,V} = \frac{1}{V} \int_V dx\, e^{iq \cdot x} a(x) = \frac{1}{\sqrt{V}} a_q$$

We learned that, because of the homogeneity of the given state ω, this operator converges strongly to an operator α_q on the representation space $\mathscr{H}$ [26]. We denote its limit by

$$\alpha_q = \lim_V \alpha_{q,V} \; ; \; \alpha_q^* = \lim_V \alpha_{q,V}^* \tag{4.12}$$

Because the $\alpha_{q,V}$ are local averages, α_q is a limit *average operator*, also sometimes called an *observable at infinity* [97], and therefore commutes with all operators of $\mathfrak{A}$, as well as with all elements of the weak closure $\mathfrak{A}''$ of the set $\mathfrak{A}$. Formulated more technically, the operator α_q is affiliated to the center $\mathscr{C} = \mathfrak{A}'' \cap \mathfrak{A}'$ of the algebra $\mathfrak{A}''$, where $\mathfrak{A}'$ is the set of bounded operators commuting with the operators $\mathfrak{A}$. Note also that we used the notation $\mathfrak{A}'' = (\mathfrak{A}')'$ used.

Let again τ_y with $y \in \mathbb{R}^d$ be the space translation canonical transformation over the distance y Eq. (7.3) of the observable algebra $\mathfrak{A}$, formally $\tau_y a^{\#}(x) = a^{\#}(x+y)$ for all creation and annihilation operators. Then we can verify readily the properties

$$\tau_y \alpha_q = e^{-iq \cdot y} \alpha_q \; ; \; \tau_y(\alpha_q^* \alpha_q) = \alpha_q^* \alpha_q \tag{4.13}$$

The operator α_q transforms under translations like the annihilation operator a_q. The second equality states that the density of particles in the q-mode, given by $n_q = \alpha_q^* \alpha_q$, is translation invariant.

From now on we assume that the state ω shows *condensation in the mode q*, or that

$$\rho_q = \omega(\alpha_q^* \alpha_q) = \omega(n_q) > 0 \tag{4.14}$$

This implies in particular that the density operator n_q is not a trivial operator, $n_q \neq 0$, and that also

$$\omega(n_q^{1/2}) \neq 0 \qquad (4.15)$$

because $\omega(n_q^{1/2}) = 0$ would be in contradiction with the separating character of each equilibrium state [26], a property expressing that if $\omega(A^*A) > 0$ for some observable A, implies that $A \neq 0$. For more technical details see [52].

Consider now the polar decomposition of the operator α_q given by

$$\alpha_q = U_q n_q^{1/2} \qquad (4.16)$$

By explicit checking α_q is a normal operator, which means that $\alpha_q^* \alpha_q = \alpha_q \alpha_q^*$. By [41], p.935, the operator U_q can be taken unitary, namely the unitary extension of the partial isometry defined by the polar decomposition. Note also that the following equality holds:

$$0 \neq \omega(n_q^{1/2}) = \omega(U_q^* \alpha_q) = \lim_V \omega(U_q^* \alpha_{q,\Lambda})$$

Therefore we can conclude that there exist a test function $h \in H = L^2(\mathbb{R}^d)$ such that

$$\omega(U_q^* a(h)) \neq 0 \qquad (4.17)$$

Note that all this holds for $q = 0$ as well as for $q \neq 0$. From now on we consider the two cases separately.

Case for $q = 0$

As mentioned, the case $q = 0$, that is, the ground state condensation, is the best known situation. It concerns homogeneous gauge invariant closed boson systems. In this case only the spontaneous gauge symmetry breaking will accompany the condensation. As it is simpler than the general case, including the case $q \neq 0$, we construct first the new gauge breaking equilibrium states in the case $q = 0$ under the condensation assumption $\rho_0 > 0$.

In this case, we define a new representation of the observable algebra acting on the GNS-representation space $\mathscr{H}$ defined by the original given equilibrium state ω. For any test function $h \in \mathscr{S}$, we define the new boson field obtained after application of the special canonical transformation Eq. (7.3) η, which is given by

$$\eta(\phi(h)) \equiv \widetilde{\phi}(h) = (U_0^* a(h) + U_0 a^*(h))^* \qquad (4.18)$$

where the unitary U_0 is given by Eq. (4.16). By definition this new field is an essentially self-adjoint operator acting on its Hilbert space $\mathscr{H}$ ([26], Lemma 5.4.12). The new boson fields $\widetilde{\phi}(h)$ define the new Weyl operators $\widetilde{W}(h) = e^{i\widetilde{\phi}(h)}$. Due to the unitarity of the operator U_0, it is trivial to check that $\widetilde{W}$ is again a representation of the observable algebra $\mathfrak{A}$ on the representation space $\mathscr{H}$ of the given state ω.

Take also the group of canonical gauge transformations $\{\tau_\varphi | \varphi \in [0, 2\pi]\}$ with $\tau_\varphi(W(h)) = W(e^{i\varphi}h)$. Define the new states ω_φ on the Weyl operators as follows:

$$\omega_\varphi(W(h)) = \omega(\widetilde{W}(e^{i\varphi}h)) \qquad (4.19)$$

For each real number φ we defined new fields $\widetilde{\phi}_\varphi(h)$ and/or new creation and annihilation operators $\widetilde{a}_\varphi(h)$ obtained from the original fields after having applied the canonical transformations, which are given by

$$\widetilde{\phi}_\varphi(h) = (U_0^* a(e^{i\varphi}h) + U_0 a^*(e^{i\varphi}h))^* \widetilde{a}_\varphi(h) = e^{-i\varphi}U_0^* a(h) = e^{-i\varphi}\eta(a(h)) \quad (4.20)$$

with η the canonical transformation defined above.

As we assumed that there is ground state condensation ($\rho_0 > 0$) it follows immediately from Eq. (4.17) that all new states ω_φ for all φ are not gauge invariant because indeed for some h, $\omega_\varphi(a(h)) = e^{i\varphi}\omega(U_0^* a(h)) \neq 0$. Note also that this expression leads immediately to the result that $\omega_\varphi \neq \omega_{\varphi'}$ if $\varphi \neq \varphi'$. Moreover by construction we get $\omega_\varphi \circ \tau_{\varphi'} = \omega_{\varphi+\varphi'}$. The gauge canonical transformation $\tau_{\varphi'}$ maps any state ω_φ into the state $\omega_{\varphi+\varphi'}$.

Furthermore, for all pairs (n,m) of natural numbers and for all test functions f we can compute the equality

$$\frac{1}{2\pi}\int_0^{2\pi} d\varphi \, \omega_\varphi(a^*(f)^n a(f)^m) = \frac{1}{2\pi}\int_0^{2\pi} d\varphi \, e^{i\varphi(m-n)}\omega(U_0^{m-n}a^*(f)^n a(f)^m)$$

$$= \delta_{n,m}\omega(a^*(f)^n a(f)^m) = \omega(a^*(f)^n a(f)^m)$$

where the last equality follows directly from the gauge invariance of the given state ω. Hence this formula proves that

$$\omega = \frac{1}{2\pi}\int_0^{2\pi} d\varphi \, \omega_\varphi \quad (4.21)$$

or that the given equilibrium state ω is equal to the integral over the set of gauge breaking states $\{\omega_\varphi \mid \varphi \in [0,2\pi]\}$ constructed above.

Next we prove that all these new states ω_φ are equilibrium states. Indeed on the basis of the equilibrium criterion Eq. (3.7) being satisfied for the starting state ω, the fact that the Hamiltonian and the number operator are gauge invariant, we get that the new states satisfy the energy-entropy criteria for equilibrium as well. Explicitly,

$$\lim_V \beta \omega_\varphi(X^*[H_V - \mu N_V, X]) = \lim_V \beta \omega(\tau_\varphi \eta(X^*)[H_V - \mu N_V, \tau_\varphi \eta(X)])$$

$$\geq \omega_\varphi(X^*X)\ln \frac{\omega_\varphi(X^*X)}{\omega_\varphi(XX^*)}$$

Finally we prove the ergodicity of all the constructed states $\{\omega_\varphi\}$. Directly from the definition of the gauge breaking states, we get using the polar decomposition Eq. (4.16), $\alpha_0 = U_0\sqrt{n_0}$, that

$$|\omega_\varphi(\alpha_0)|^2 = \omega(\sqrt{n_0})^2 = \omega_\varphi(\sqrt{n_0})^2.$$

Using this equality and taking into account that α_0 is a normal operator (implying that U_0 and $\sqrt{n_0}$ commute) and working within the GNS-construction Eq. (7.1) of the state $\omega_\varphi(.) = (\Omega_\varphi, . \Omega_\varphi)$ transforms the following Schwartz inequality into an equality. Indeed we get

$$\omega(\sqrt{n_0})^2 = |\omega_\varphi(\alpha_0)|^2 = |(\Omega_\varphi, U_0\sqrt{n_0}\Omega_\varphi)|^2$$
$$= |(\sqrt[4]{n_0}\Omega_\varphi, U_0\sqrt[4]{n_0}\Omega_\varphi)|^2 \le ||\sqrt[4]{n_0}\Omega_\varphi||^2 = \omega(\sqrt{n_0})^2$$

Hence the two vectors, Ω_φ and $U_0\sqrt{n_0}\Omega_\varphi$ are proportional vectors, that is, there exists a complex number κ such that $\kappa\Omega_\varphi = U_0\sqrt{n_0}\Omega_\varphi = \alpha_0\Omega_\varphi$. We readily obtain $|\omega_\varphi(\alpha_0)|^2 = |(\Omega_\varphi, \alpha_0\Omega_\varphi)|^2 = |\kappa|^2 = (\alpha_0\Omega_\varphi, \alpha_0\Omega_\varphi) = \omega_\varphi(n_0) = \omega(n_0) = \rho_0$, proving the ergodicity property of the states ω_φ expressed in Eq. (4.11), as well as their property of showing off-diagonal long range order Eq. (2.33).

All these properties together prove the first part of the following theorem:

Theorem 4.2. *With the definitions and notations of above, if there is zero-mode condensation ($\rho_0 > 0$) for the limit Gibbs or any other equilibrium state ω, satisfying the conditions A and B, then there exists a set of ergodic equilibrium states ω_φ Eq. (4.19), all breaking the gauge symmetry, whose convex combination with integral over the phase φ yields the original equilibrium state. Following trivially from Schwartz inequality we already sketched, the converse of this statement holds as well: If there exists any gauge breaking equilibrium state of the type discussed above, then there is ground state condensation.*

We should realize the constructive nature of the proof of this theorem, where the gauge breaking equilibrium states ω_φ, (see Eq. (4.19)) are explicitly constructed in terms of any given thermodynamic limit Gibbs state ω with a non-vanishing condensate density $\rho_0 > 0$. The mathematical proof we gave may come across as somewhat technical, difficult, and nontransparent to those less mathematically inclined.

Being especially concerned with the more physics-minded among us, it may be instructive to give a direct formal expression of the gauge breaking states ω_φ in terms of the correlation functions. This form of the states may come across as more suggestive or understandable solely because they are written in terms of the creation and annihilation operators and the condensate density. Following exactly the construction procedure sketched above with the same notations and definitions, in particular for any limit Gibbs state ω,

$$\omega(X) = \lim_V \omega_V(X) \equiv \lim_V \frac{Tr\, e^{\beta(H_V - \mu N_V)} X}{Tr\, e^{\beta(H_V - \mu N_V)}}$$

where X is any local observable with fixed density

$$\rho = \lim_V \omega_V\left(\frac{N_V}{V}\right)$$

and with a nonzero condensate density $\rho_0 = \omega(n_0) > 0$, we obtain each of the gauge breaking states, determined by the (n,m)-correlation functions, for each value of the phase φ, by the following: For any creation/annihilation operator $a^*(f), a(f)$ we find

$$\omega_\varphi(a^*(f)^n a(f)^m) = e^{i(m-n)\varphi} \lim_V \omega\left(\frac{a_0^n (a_0^*)^m}{(a_0^* a_0)^{(n+m)/2}} a^*(f)^n a(f)^m\right) \qquad (4.22)$$

Expressed explicitly in terms of the limit Gibbs state one gets

$$\omega_\varphi(a^*(f)^n a(f)^m) = e^{i(m-n)\varphi} \lim_V \frac{tr\, e^{-\beta(H_V - \mu N_V)} \frac{a_0^n (a_0^*)^m}{(a_0^* a_0)^{(n+m)/2}} a^*(f)^n a(f)^m}{tr\, e^{-\beta(H_V - \mu N_V)}}$$

This explicit form of the equation is of genuine importance in the theory of Bose-Einstein condensation. The result shows explicitly the intimate link between condensation and spontaneous symmetry breaking. It shows also the link between the original equilibrium state and the gauge breaking states. Equation (4.22) might come across as simple. The above mathematic computation merely demonstrates one way of showing that the previous construction is mathematically meaningful and physically realistic. In any case it serves as a better understanding of the structural aspects of the links between the phenomena of spontaneous gauge symmetry breaking, of off-diagonal long range order, and of ground state condensation for equilibrium states of boson systems.

Case for $q \neq 0$

Now we turn to the case of non-ground state condensation. This means that we start with an equilibrium state ω with the property of showing condensation $\rho_q > 0$ for some $q \neq 0$. Again we define a new representation of the observable algebra $\mathfrak{A}$ acting on the representation space $\mathscr{H}$ of the equilibrium state. For any test function $h \in \mathscr{S}$, we again define the new boson field by

$$\tilde{\phi}(h) = (U_q^* a(h)) + U_q a^*(h))^*$$

where U_q is again the unitary operator of the polar decomposition Eq. (4.16). For notational convenience the q-dependence of the new field $\tilde{\phi}$ is not explicitly indicated. Again the field by definition is an essentially self-adjoint operator acting on $\mathscr{H}$ and the new Weyl operator map $\tilde{W}$ maps once again $\mathscr{S} \rightarrow \mathscr{B}(\mathscr{H})$, the bounded operators acting on $\mathscr{H}$,

$$\tilde{W}(h) = e^{i\tilde{\phi}(h)} , \; h \in \mathbb{R}^d \tag{4.23}$$

Because the U_q are unitary operators commuting with all observables, we can check that $\tilde{W}$ is a representation of the observable algebra $\mathfrak{A}$ on the Hilbert space $\mathscr{H}$ (comparable to Eq. (4.18)). Let us denote again by η the canonical transformation map of $\mathfrak{A}$ into the bounded operator algebra $\mathscr{B}(\mathscr{H})$ of operators acting on $\mathscr{H}$

$$\eta(W(h)) = \tilde{W}(h) \tag{4.24}$$

From this Weyl operator relation, we immediately obtain the action of η on the creation and annihilation operators:

$$\tilde{a}(h) \equiv \eta(a(h)) = U_q^* a(h) \; ; \; \tilde{a}^*(h) \equiv \eta(a^*(h)) = U_q a^*(h)$$

So far all this is comparable to the $q = 0$ case. The space translations now enter the discussion. Again for any equilibrium state ω and any $x \in \mathbb{R}^d$, we define the functional $\tilde{\omega}_x$ on the boson algebra $\mathfrak{A}$ by

$$\tilde{\omega}_x(W(h)) = (\Omega, \tilde{W}(h_x)\Omega) = \omega(\eta \cdot \tau_x(W(h))) \tag{4.25}$$

where $h_x(y) = h(x+y)$ is again the function h translated over the distance x. Clearly for each x, $\tilde{\omega}_x$ is again a state of the boson algebra.

As above, the states $\tilde{\omega}_x$, are not gauge invariant, a property following directly from the definition in Eq. (4.25). In particular, we also have that $\tilde{\omega}_x(a(h)) = \omega(U_q^* \tau_x a(h)) \neq 0$ for some $h \in H$, which is proved in Eq. (4.17). Also directly from their very definition, the states $\tilde{\omega}_x$ satisfy the relation: For all $x, y \in \mathbb{R}^d$,

$$\tilde{\omega}_x \cdot \tau_y = \tilde{\omega}_{x+y} \tag{4.26}$$

The translation canonical transformations τ_y map the state $\tilde{\omega}_x$ onto the state $\tilde{\omega}_x \cdot \tau_y = \tilde{\omega}_{x+y}$. The set of states of the type defined in Eq. (4.25) is generated by applying the translation canonical transformations on any of them. The question is: Are all these states different states? In other words, how large is this set of states which we generated in this way?

We analyze these questions and focus ourselves on the different directions of the translations. Let $q = (2\pi/\gamma)e$, with $e \in \mathbb{R}^d$ the unit vector in the direction of q and γ the period or wavelength of the mode q, or $\gamma = 2\pi/|q|$. We denote by $e_\perp$ any unit vector orthogonal to e. We therefore obtain the first property:

Lemma 4.3. *With the definitions and notations introduced above, all states $\tilde{\omega}_x$, $x \in \mathbb{R}^d$ are constructed from the equilibrium state ω (see Eq. (4.25)) and satisfy:*

1. Each state $\tilde{\omega}_x$ is space translation invariant in any direction $e_\perp$ orthogonal to q. In particular $\forall t \in \mathbb{R}$, we obtain the equality

$$\tilde{\omega}_{x+te_\perp} = \tilde{\omega}_x$$

2. Each state $\tilde{\omega}_x$ is periodic with period γ in the direction of q expressed by

$$\tilde{\omega}_{x+\gamma e} = \tilde{\omega}_x$$

Proof. Using the GNS-representation Eq. (7.1), we can write the states in the form

$$\tilde{\omega}_{x+y}(W(h)) = (\Omega, e^{i\tilde{\phi}(h_{x+y})}\Omega)$$

yielding for the corresponding fields

$$\tilde{\phi}(h_{x+y}) = (U_q^* \tau_y a(h_x) + U_q \tau_y a^*(h_x))^* = \tau_y(\tau_{-y}(U_q^*)a(h_x) + \tau_{-y}(U_q)a^*(h_x))^*$$

Using the action of the translations on the unitary operators Eq. (4.13) we get

$$\tau_y(U_q) = U_q e^{-iq \cdot y}$$

Hence for $y = te_\perp$ and, as $q \cdot \gamma e = 2\pi$, for any $y = \gamma e$ we get the translation invariance of U_q or $\tau_y(U_q) = U_q$. Hence in both of these cases

$$\tilde{\phi}(h_{x+y}) = \tau_y \tilde{\phi}(h_x)$$

Using the space translation invariance (condition A) of the given state ω, the proofs of Items *1* and *2* follow immediately.

The Lemma shows that the new states $\widetilde{\omega}_x$ are periodic states with period γ in the q-direction and that they remain translation invariant in any direction orthogonal to the vector q. Now we concentrate on the states and their translations in the q-direction within the period or within the interval $[0, \gamma]$.

Lemma 4.4. *1. For all $t, t' \in [0, \gamma]$ with $t \neq t'$, we get $\widetilde{\omega}_{te} \neq \widetilde{\omega}_{t'e}$, or the two states differ from each other.*

2. The originally given state ω is equal to the convex combination with equal weight of all, the two by two different, non-homogeneous gauge breaking states $\{\widetilde{\omega}_{te} | t \in [0, \gamma]\}$. We therefore obtain the equality of states

$$\omega = \frac{1}{\gamma} \int_0^\gamma dt \, \widetilde{\omega}_{te}$$

3. All states in the set $\{\widetilde{\omega}_{te} | t \in [0, \gamma]\}$ are ergodic equilibrium states showing the same q-mode condensation.

Proof. As the state ω is analytic (condition B) it follows from their definition that all the states $\widetilde{\omega}_x$ are also analytic. Therefore it is again sufficient to prove the statements of the theorem on the monomials in the field operators (see also [26], p.38) or, in other words, for all (n, m)-correlation functions of our states. Consider first the one-point function

$$\widetilde{\omega}_{te}(a(h)) = \omega(U_q^* \tau_{te} a(h))$$

A straightforward computation using the space translation invariance of the given state ω yields

$$\widetilde{\omega}_{te}(a(h)) = e^{-iq \cdot te} \omega(U_q^* a(h)) = e^{-i\frac{2\pi}{\gamma}t} \omega(U_q^* a(h))$$

As there exists an element $h \in H$ such that $\omega(U_q^* a(h)) \neq 0$ Eq. (4.17), and as for $t, t' \in [0, \gamma], t \neq t'$ holds that

$$(\widetilde{\omega}_{te} - \widetilde{\omega}_{t'e})(a(h)) = \left(e^{-it\frac{2\pi}{\gamma}} - e^{-it'\frac{2\pi}{\gamma}} \right) \omega(U_q^* a(h)) \neq 0$$

proving item *1* of the Lemma.

To prove the statement in item 2, consider for all natural numbers m and n, and for any test function f, the integrals

$$\frac{1}{\gamma} \int_0^\gamma dt \, \widetilde{\omega}_{te}(a^*(f)^n a(f)^m) = \frac{1}{\gamma} \int_0^\gamma dt \, \omega(\tau_{te}(a^*(f)^n) \tau_{te}(a_{te}(f)^m) U_q^n U_q^{-m})$$

Space translation invariance of the state ω (condition A) yields

$$\frac{1}{\gamma} \int_0^\gamma dt \, \widetilde{\omega}_{te}(a^*(f)^n a(f)^m)$$
$$= \frac{1}{\gamma} \int_0^\gamma dt \, e^{i\frac{2\pi}{\gamma}t(m-n)} \omega(a^*(f)^n a(f)^m U_q^{n-m})$$
$$= \delta_{n,m} \omega(a^*(f)^n a(f)^m) = \omega(a^*(f)^n a(f)^m)$$

proving the statement in item 2.

Now we show that for any t in the interval $[0, \gamma]$, any state $\tilde{\omega}_{te}$ is an equilibrium state by showing that these states again satisfy the equilibrium conditions Eq. (3.7). Let X be an arbitrary monomial in the creation and annihilation operators, using condition A for the state ω together with the space translation and gauge invariance of the Hamiltonian Eq. (2.4), we obtain the EEB-inequalities

$$
\begin{aligned}
&\beta \lim_V \tilde{\omega}_{te}(X^*[H_V - \mu N_V, X]) \\
&= \beta \lim_V \omega(\eta \circ \tau_{te}(X^*[H_V - \mu N_V, X]) \\
&= \beta \lim_V \omega((\eta \circ \tau_{te}(X))^*[H_V - \mu N_V, \eta \circ \tau_{te}(X)]) \\
&\geq \omega((\eta \circ \tau_{te}(X))^*(\eta \circ \tau_{te}(X))) \ln \frac{\omega((\eta \circ \tau_{te}(X))^* \eta \circ \tau_{te}(X))}{\omega((\eta \circ \tau_{te}(X))(\eta \circ \tau_{te}(X))^*)} \\
&= \tilde{\omega}_{te}(X^*X) \ln \frac{\tilde{\omega}_{te}(X^*X)}{\tilde{\omega}(XX^*)}
\end{aligned}
$$

which shows indeed that all the newly constructed states satisfy the energy-entropy criteria for equilibrium and therefore are equilibrium states.

Finally the ergodicity property of the constructed states follows the same argument as that used for ground state condensation (see proof of the theorem Eq. (4.2)).

This Lemma proves that if we have an equilibrium state ω showing condensation in the q-mode Eq. (4.10) then there exists a set of periodic ergodic equilibrium states $\{\omega_{te} | t \in [0, \gamma]\}$ in the q-direction with period γ. In directions orthogonal to q, these states remain space homogenous. Note that these states break at the same time the gauge symmetry. As an interesting technical property we observe that all these states live in the same representation space $\mathcal{H}$, namely the representation space of the observable algebra $\mathfrak{A}$ determined by the original equilibrium state ω. All these properties yield a better understanding of the relations between spontaneous space translation, gauge symmetry breaking, space translation symmetry breaking, and boson q-mode condensation. In particular we obtain a better view of the relative position of the different equilibrium states, which can be ergodic or non-ergodic.

Above we analyzed the situation for one single direction q. It is a student exercise to generalize the analysis to an arbitrary number of dimensions. Indeed suppose that we have more than one condensation mode. Suppose for instance that we have d such modes characterized by the d linear independent vectors $q_1, \ldots, q_d$. This means that we start with an equilibrium state ω, which has the condensate densities $\rho_{q_i} > 0$ for $i = 1, \ldots d$. All vectors are taken independent and therefore we may consider the case that $(q_i, q_j) = \delta_{ij}|q_i|^2$. We can perform the above constructions in each direction independently because, in any case, the corresponding α_{q_i}-operators, relevant in the proofs, commute with each other and with their adjoint operators. We obtain equilibrium states $\tilde{\omega}_x$, with $x = t_1 e_1 + \cdots + t_d e_d$, which are periodic in all directions with in general d different periods γ_i. These different directions are creating a discrete d-dimensional lattice symmetry group for these new equilibrium states. The spontaneous symmetry breaking is breaking the full symmetry group $\mathbb{R}^d$ down to a lattice group. The occurrence of such lattice groups has been discussed in experiments with trapped boson systems (see [140] and references therein).

General Formulation

Finally we recapitulate what we proved as general results about the one-to-one relation between Bose-Einstein condensation and spontaneous symmetry breaking as the following theorem:

Theorem 4.5. *On the basis of what was proven above, any homogeneous gauge invariant limit Gibbs state* ω *showing condensation* $(\rho_q > 0)$ *in one or more modes* q, *can be written as the integral of the equilibrium states, if* $q = 0$, *Eq. (4.19),*

$$\{\widetilde{\omega}_\varphi | \varphi \in [0, 2\pi]\}$$

if $q \neq 0$, *Eq. (4.25),*

$$\{\widetilde{\omega}_x | x \in [0, \gamma]; \gamma = \frac{2\pi}{|q|}\}$$

with the following properties:

1. *All these states are explicitly constructed from the given state* ω.
2. *All states break the gauge symmetry because* $\widetilde{\omega}_\varphi(\alpha_0) \neq 0$ *and* $\widetilde{\omega}_x(\alpha_q) \neq 0$. *If* $q \neq 0$ *the states* $\widetilde{\omega}_x$ *break the translation symmetry and are periodic in the* q-*direction with period* γ.
3. *All states are ergodic equilibrium states with* q-*mode condensation, and are showing off-diagonal long range order Eq. (2.33)*

$$\lim_V \{|\widetilde{\omega}_\varphi(\frac{a_0}{\sqrt{V}})|^2 - \widetilde{\omega}_\varphi(\frac{a_0^* a_0}{V})\} = \lim_V |\widetilde{\omega}_\varphi(\frac{a_0}{\sqrt{V}})|^2 - \rho_0 = 0$$

$$\lim_V \{|\widetilde{\omega}_x(\frac{a_q}{\sqrt{V}})|^2 - \widetilde{\omega}_x(\frac{a_q^* a_q}{V})\} = \lim_V |\widetilde{\omega}_x(\frac{a_q}{\sqrt{V}})|^2 - \rho_q = 0$$

A converse statement holds as well: If gauge breaking periodic states appear in some direction q, *then there is* q-*condensation.*

Proof. The first statement of the Theorem follows directly from the two preceding Lemmas.

The converse statement is an immediate consequence of the Schwartz inequality $0 < |\omega(a_q)|^2 \leq \omega(a_q^* a_q)$ and the periodicity of the state.

In the case of ground state $(q = 0)$-condensation we wrote down a formal but, as physicists might consider, an intuitive and understandable expression Eq. (4.22) for the symmetry breaking states explicitly in terms of the correlation functions of the given limit Gibbs state. For those inclined, an interesting exercise is to write the analogous expression for the case of q-condensation.

We should note that in both cases discussed above a proof is given of the property that, if there is Bose condensation, then the limit Gibbs states ω_β are non-trivial convex combinations of the symmetry breaking equilibrium states $\{\widetilde{\omega}_\varphi\}_\varphi$ or $\{\widetilde{\omega}_{te}\}_t$. The following have previously been derived:

$$\omega_\beta = \frac{1}{2\pi} \int_0^{2\pi} d\varphi \, \tilde{\omega}_\varphi \text{ and } \omega_\beta = \frac{1}{\gamma} \int_0^\gamma dt \, \tilde{\omega}_{te} \tag{4.27}$$

In particular these equations express that in the condensation region the limit Gibbs state ω_β is no longer an ergodic state Eq. (2.26). At the contrary it is a non-trivial convex composition of ergodic states.

Whenever we are interested in the equilibrium states of a concrete Bose system, hence in the solutions of the equilibrium conditions, it should come over as natural to always look for the ergodic or space extremal states. In the condensation region, they coincide with the states $\{\tilde{\omega}_\varphi\}_\varphi$ or $\{\tilde{\omega}_{te}\}$, not with the limit Gibbs state ω_β.

Ergodic states have the interesting property that space averages of local observables are multiples of the identity. This is an immediate consequence of the property Eq. (2.27). Applying this property for the special ergodic states $\tilde{\omega}_\varphi$ or $\tilde{\omega}_{te}$, the average operators α_0 or α_q Eq. (4.12) extensively used in the proof of the Theorem Eq. (4.5) are multiples of the unit operator, or are essentially what we sometimes call in this field *c-numbers*.

$$c = \tilde{\omega}_\varphi(\alpha_0) = \sqrt{\rho_0} e^{i\varphi} \text{ and } c = \tilde{\omega}_{te}(\alpha_q) = \sqrt{\rho_q} e^{i\varphi_{q,t}} \tag{4.28}$$

These numbers refer to the so-called c-numbers of the c-number assumption in the theory of Bogoliubov which is expressed by

$$c = \alpha_0 \; ; \; c = \alpha_q$$

In the context of this c-number philosophy, the phases φ and $\varphi_{q,t}$ are sometimes called the *phases of the states* $\tilde{\omega}_\varphi$ and $\tilde{\omega}_{te}$. Therefore Theorem Eq. (4.5) might give the impression of having stepped toward understanding the position of what is called the *Bogoliubov approximation*. Let us remark however that originally, but still highly popular, this approximation consists in substituting the condensate creation and annihilation operators $\alpha_{q,V}^\# = a_q^\#/\sqrt{V}$ by c-numbers in the Hamiltonian, which is however an operation of a totally different order. Indeed making this kind of approximation means that we are neglecting the quantum fluctuations of the creation and annihilation operators of this q-mode operators (see Eq. (6)), which remains a rather drastic approximation. In the literature there have been many attempts to prove that this Bogoliubov approximation is not an approximation but an exact statement. A first serious attempt towards such a proof was on the level of a variational principle of the energy density as a function of the condensate density. This attempt is found in [57]. On the other hand, it is clear that the theorem Eq. (4.5) does not contribute to unraveling the question about the exactness of this approximation. In fact we did not use nor touch this approximation. Later, at the occasion of a more explicit discussion of the theory of Bogoliubov, we come back to this point and we make more precise statements about the position concerning this issue.

Spontaneous Symmetry Breaking in Classical Lattice Systems

To illustrate the universality, reaching far beyond boson systems, of the ideas and the techniques behind the analysis described above concerning the relation between condensation in boson systems and spontaneous symmetry breaking and off-diagonal long-range order, we formulate and prove the equivalent result for classical spin lattice systems. In particular we give the construction of symmetry breaking states for these classical equilibrium systems if and only if there is long-range order. The content of this note can also be found in [163].

We start with the definition of the classical spin system which consists again in the specification of its algebra of observables and its set of states. The set of observables of a classical spin system on a d-dimensional square lattice $\mathbb{Z}^d$, consists of the commutative algebra $\mathfrak{A}$ generated by the one-site observables $\{\sigma_x / x \in \mathbb{Z}^d\}$. In other words each system observable, denoted by X or $X(\sigma)$, is of the type

$$\sum_n \sum_{x_1,\dots,x_n} c_{x_1,\dots,x_n} \sigma_{x_1} \sigma_{x_2} \dots \sigma_{x_n}$$

where $x_1,\dots,x_n \in \mathbb{Z}^d$ and the $c_{x_1,\dots,x_n}$ are complex numbers. The spin variables σ_x are functions taking the values ± 1.

Homogeneous spin systems are defined by local Hamiltonians H_Λ, one for each finite subset Λ of the lattice, of the form:

$$H_\Lambda = \Sigma_{\Delta \subseteq \Lambda} \, \phi(\Delta) \sigma_\Delta \tag{4.29}$$

where we used the notation $\sigma_\Delta = \prod_{x \in \Delta} \sigma_x$. The translation invariance is guaranteed by the interaction energy condition $\phi(\Delta + a) = \phi(\Delta)$ holding for all lattice translations a and subsets Δ of $\mathbb{Z}^d$. Furthermore a condition on the potential ϕ is applied in order to guarantee that the Hamiltonian describes a normal extensive system. We do not enter here in more details.

The *global spin flip operation* Θ maps each of the spin variables σ_x onto its spin-flipped opposite $-\sigma_x$.

For Λ any finite subset of the lattice points, we denote by Θ_Λ the *local spin flip operation* of all spins σ_x with x in Λ. Not only translation invariance of the systems is imposed, we assume also the *spin flip invariance* of our systems: i.e. we assume that all local Hamiltonians satisfy the condition $\Theta(H_\Lambda) = H_\Lambda$ for all $\Lambda \subset \mathbb{Z}^d$. We notice that in this case we deal with a discrete symmetry group.

Clearly the best known prototype model systems are the d-dimensional Ising models $H_\Lambda = -J\sum_{<x,y>;x,y \in \Lambda} \sigma_x \sigma_y$, where $<x,y>$ stand for the nearest neighbor sites x and y. Of course the Hamiltonians formulated in Eq. (4.29) contain many more classical spin models far beyond the Ising models.

We are again interested in the equilibrium states, which are expectation-valued maps or for these classical systems probability measures on the set of functions $\mathfrak{A}$ the observables of these systems.

First of all we can again consider the limit *Gibbs states*, denoted by ω_β with β the inverse temperature, and naturally defined by

$$\omega_\beta(X(\sigma)) = \lim_{\Lambda \to \mathbb{Z}^d} \frac{\Sigma_{\{\sigma=\pm 1; \sigma \in \Lambda\}} X(\sigma) \exp\{-\beta H_\Lambda(\sigma)\}}{\Sigma_{\{\sigma=\pm 1; \sigma \in \Lambda\}} \exp\{-\beta H_\Lambda(\sigma)\}} \tag{4.30}$$

There are many possible thermodynamic limits $\Lambda \to \mathbb{Z}^d$ which can be considered. Again the limits can depend on the geometrical forms of the sequences of Λ's. Different sequences can yield possibly different limit Gibbs states. By definition, each of these limit Gibbs states, is denoted by the same symbol ω_β and is homogeneous and spin flip invariant.

Analogous to quantum systems, any state ω of the spin system is an *equilibrium state* of the system if it satisfies the classical *Energy-Entropy Balance (EEB) criterion at* β, that is, if for each fixed finite lattice subset Λ and any non-negative observable $X \geq 0$,

$$\lim_{\Lambda} \omega(X(\sigma)(\Theta_{\tilde{\Lambda}} H_\Lambda(\sigma) - H_\Lambda(\sigma))) \geq \frac{1}{\beta} \omega(X(\sigma)) \ln \frac{\omega(X(\sigma))}{\omega(\Theta_{\tilde{\Lambda}}(X(\sigma)))} \qquad (4.31)$$

As in the quantum case, these conditions are handy tools as criteria for equilibrium states (probability measures), they are nothing but the set of Euler equations for the basic free-energy density functional variational principle of classical statistical mechanics of these systems. We should not be surprised by the inequalities instead of equalities. It has been proven that the system of inequalities Eq. (4.31) is equivalent to the system of Euler equation equalities of the classical variational principle (see also [55]). The latter equalities are however practically less manageable in applications than the inequalities. For all these reasons the EEB criterion holds again as firm general defining criterion for the classical spin systems equilibrium states.

Again we show [55] that each limit Gibbs state satisfies this *Energy-Entropy Balance (EEB) criterion*. Each Gibbs state ω_β is clearly homogeneous and spin-flip invariant ($\omega_\beta \circ \Theta = \omega_\beta$). But there may exist more homogeneous states ω satisfying the EEB criterion. Some of them may break the Θ-symmetry invariance of the given system $\{H_\Lambda\}$. If this happens we speak about the occurrence of *spontaneous symmetry breaking* of the spin-flip symmetry.

For any homogeneous state ω of the system, the *magnetization* of the state is given by

$$\omega(\sigma_0) = \omega(\sigma_y) = \lim_{\Lambda \to \mathbb{Z}^d} \omega\left(\frac{\sum_{x \in \Lambda} \sigma_x}{|\Lambda|}\right) \qquad (4.32)$$

where y is any arbitrary lattice point and where $|\Lambda|$ stands for the volume or the number of lattice points of Λ.

We can check that averages of local observables(functions), say A, again always exist within the GNS-representation of a homogeneous state. The Hilbert space $\mathscr{H}$ is the closure of the set $\mathfrak{A}$ with respect to the scalar product $(X,Y) \equiv (X\Omega_\omega, Y\Omega_\omega) \equiv \omega(X^*Y)$, where $\Omega_\omega = \mathbf{1}$ stands for the the unit observable. With this in mind, if τ_a denotes the translation action $\tau_a(\sigma_x) = \sigma_{x+a}$ over the distance a, then for all observables X, Y we obtain

$$\omega(X\overline{A}Y) = \lim_{\Lambda} \omega\left(X\left(\frac{1}{|\Lambda|} \sum_{a \in \Lambda} \tau_a A\right) Y\right) \qquad (4.33)$$

defining the *average observable* $\overline{A} \equiv \lim_{\Lambda} \frac{1}{|\Lambda|} \sum_{a \in \Lambda} \tau_a A$. We now set $A = \sigma_x$ for some point x. Then $\overline{A} = \overline{\sigma_x} \equiv \overline{\sigma}$ and Eq. (4.32) becomes

$$\omega(\sigma_y) = \omega(\overline{\sigma}) \tag{4.34}$$

By the Schwartz inequality we get

$$\omega(\sigma_y)^2 = \omega(\overline{\sigma})^2 \leq \omega(\overline{\sigma^2}) \tag{4.35}$$

expressing the following property: If ω is a homogeneous state breaking the symmetry (i.e., if $\omega(\sigma_y) \neq 0$), then $\omega(\overline{\sigma^2}) > 0$.

Clearly any state ω with the property $\omega(\overline{\sigma^2}) > 0$ may be called in a natural sense a *state with macroscopic occupation of spin density*. In the rest of this note we prove the following statement:

STATEMENT *If we have a limit Gibbs state ω_β showing a macroscopic occupation of spin density, hence with the property $\omega_\beta(\overline{\sigma^2}) > 0$, then there exists a spin symmetry breaking equilibrium (i.e. satisfying Eq. (4.31)) states ω_+ and ω_-, which satisfy moreover the equality*

$$\omega(\sigma_y)^2_\pm = \omega_\pm(\overline{\sigma})^2 = \omega_\pm(\overline{\sigma^2}) \tag{4.36}$$

that is, they satisfy Eq. (4.35) but with the equality sign. Moreover the limit Gibbs state is an equal weight convex combination of the states $\omega_\pm$:

$$\omega_\beta = \frac{1}{2}\omega_+ + \frac{1}{2}\omega_-$$

The implications of this equality sign case are the following: If we have spontaneous symmetry breaking states $\omega_\pm$ and the equality sign in Eq. (4.36), then these states have the property of showing the long-range order property

$$\omega_\pm(\overline{\sigma^2}) = |\omega_\pm(\overline{\sigma})|^2 > 0 \tag{4.37}$$

This property is similar to the notion of "off-diagonal long range order" (see [129]), a notion which has been introduced in the context of quantized fields.

The statement expresses also that, if we have a limit Gibbs state ω_β showing a macroscopic occupation of spin density, then the symmetry breaking states $\omega_\pm$ always exist and they show long-range order.

Construction of the states $\omega_\pm$.

We start from the given limit Gibbs state ω_β, which is homogeneous, spin-flip invariant, and satisfies the property $\omega_\beta(\overline{\sigma^2}) > 0$. In particular, spin-flip invariance implies that $\omega_\beta(\sigma_{x_1}...\sigma_{x_{2n+1}}) = 0$ for all integers n. Consider the average spin function $\overline{\sigma}$ Eq. (4.33) in the representation induced by the state ω_β. This average is a real function with values in the interval $[-1, 1]$. Consider the polar decomposition of this average function

$$\overline{\sigma} = U\sqrt{\overline{\sigma^2}} \tag{4.38}$$

As $\omega_\beta(\overline{\sigma^2}) > 0$, we have $\overline{\sigma} \neq 0$ or $\overline{\sigma}$ is a non-trivial function in this representation. The function U is real function taking the values ± 1 or $U^2 = 1$ on the support of $\overline{\sigma}$. We can also write $U = \overline{\sigma}/\sqrt{\overline{\sigma^2}}$. The function U can also be extend by one outside the support, such that $U^2 = 1$ everywhere.

Define the new spin variables $\tilde{\sigma}_x$ for all x in the lattice

$$\tilde{\sigma}_x = U\sigma_x \equiv \eta(\sigma_x)$$

where η is a morphism of the algebra $\mathfrak{A}$ generated by the σ's unto itself. The new variables $\tilde{\sigma}_x$ generate a new representation of the original algebra of observables. We define now the states $\omega_\pm$ as follows: For each observable X, the expectation values is given by

$$\omega_+(X) = \omega_\beta(\eta(X)); \quad \omega_-(X) = \omega_\beta(\eta(\Theta(X))) \tag{4.39}$$

Properties of the states $\omega_\pm$.

1. It is readily checked, using the definition formulae Eq. (4.39) based on the given Gibbs state ω_β, with β finite, that the states $\omega_\pm$ satisfy the EEB criterion Eq. (4.31). Therefore the newly constructed states are equilibrium states.
2. We compute that

$$\omega_\pm(\sigma_{x_1}...\sigma_{x_{2n}}) = \omega_\beta(\sigma_{x_1}...\sigma_{x_{2n}})$$

implying that the new states and the Gibbs state coincide on the even monomials in the σ's. Also, as $U^2 = 1$,

$$\omega_+(\sigma_{x_1}...\sigma_{x_{2n+1}}) = -\omega_-(\sigma_{x_1}...\sigma_{x_{2n+1}})$$

implying all together that for all observables X

$$\omega_\beta(X) = \frac{1}{2}\omega_+(X) + \frac{1}{2}\omega_-(X) \tag{4.40}$$

The given Gibbs state ω_β is written as an equal weight convex combination of the two constructed states $\omega_\pm$.
3. From the definition formulae Eq. (4.39) we compute the formulae

$$\omega_\pm(\overline{\sigma}) = \pm\omega_\beta(\sqrt{\overline{\sigma}^2}) \tag{4.41}$$

Using now the Hilbert space representations (GNS-representation [26]) of the states $\omega_\pm$, respectively ω_β:

$$\omega_\pm(X) = (\Omega_\pm, X\Omega_\pm); \quad \omega_\beta(X) = (\Omega_\beta, X\Omega_\beta)$$

As

$$\omega_\beta(\sqrt{\overline{\sigma}^2})^2 = |\omega_\pm(\overline{\sigma})|^2 = |(\Omega_\pm, U\sqrt{\overline{\sigma}^2}\Omega_\pm)|^2 = |(\sqrt[4]{\overline{\sigma}^2}\Omega_\pm, U\sqrt[4]{\overline{\sigma}^2}\Omega_\pm)|^2$$

which by the Schwartz inequality is majorized by

$$\leq (\Omega_\pm, \sqrt{\overline{\sigma}^2}\Omega_\pm)^2 = \omega_\beta(\sqrt{\overline{\sigma}^2})^2$$

This implies that the vectors $\Omega_\pm$ are proportional to the vectors $U\sqrt{\overline{\sigma}^2}\Omega_\pm$ or that there exists a complex number $\kappa_\pm$ such that

$$\kappa_{\pm}\Omega_{\pm} = U\sqrt{\overline{\sigma}^2}\Omega_{\pm} = \overline{\sigma}\Omega_{\pm}$$

We get

$$|\omega_{\pm}(\overline{\sigma})|^2 = |(\Omega_{\pm}, \overline{\sigma}\Omega_{\pm})|^2 = |\kappa_{\pm}|^2 = (\overline{\sigma}\Omega_{\pm}, \overline{\sigma}\Omega_{\pm}) = \omega_{\pm}(\overline{\sigma}^2) = \omega_{\beta}(\overline{\sigma}^2) > 0$$

The last inequality follows from the property of macroscopic occupation of spin density for the state ω_{β}. This relation proves that the states $\omega_{\pm}$ have the property of showing spontaneous symmetry breaking(SSB) as well as that of showing long-range order Eq. (4.36).

All this proves the statement which we want to make for classical spin systems. We make a number of final notes.

We may recall the canonical model independent construction Eq. (4.39) of the symmetry breaking states (SSB states) and remember the usual ways of showing the existence of spontaneous symmetry breaking states for the Ising systems, which consist of fixing plus or minus symmetry breaking boundary conditions. This means breaking the symmetry by brut force. There are also the plus combined minus boundary conditions leading to non-homogeneous states with an interface structure. Non-homogeneous states lie outside the scope of this note. In [155] we find the main results for Ising systems. The $\pm$-boundary condition technique is considered to give a physical interpretation of the origins of the symmetry breaking. However one can also consider symmetry breaking external field perturbations, which tend to zero at the end of the argument. Model computations [166] show boundary condition dependencies on the various volume rates at which this limit is taken to zero. For quantum systems but also for classical systems, boundary conditions can show a complicated picture. The construction Eq. (4.39) of the SSB-states is independent from boundary-conditions considerations. It is based solely based on the notion of macroscopic occupation of spin density for a limit Gibbs state.

Finally, we stress that our construction of SSB-states yields again an immediate and explicit relation between a limit Gibbs state (ω_{β}) and the SSB-states. For convenience we rewrite the SSB-states once more in the following form: For any $n \in \mathbb{N}$, we have

$$\omega_+(\sigma_{x_1}\ldots\sigma_{x_n}) = \omega_{\beta}((\frac{\overline{\sigma}}{\sqrt{\overline{\sigma}^2}})^n \sigma_{x_1}\ldots\sigma_{x_n})$$

$$\omega_-(\sigma_{x_1}\ldots\sigma_{x_n}) = (-1)^n \omega_{\beta}((\frac{\overline{\sigma}}{\sqrt{\overline{\sigma}^2}})^n \sigma_{x_1}\ldots\sigma_{x_n})$$

As an overall global remark concerning the material of this section we add the following: The ideas and techniques developed for the construction of symmetry breaking states can be applied to many more systems (e.g. Heisenberg models) than boson and classical spin systems. Of course we focus on quantum and classical spin models with discrete or continuous symmetries, but we can also consider field theoretic models with larger symmetry groups.

4.3.3 Condensate Equations

What is the *condensate equation*? As is by now well known, the always challenging problem of boson systems is to show that there exists one or more equilibrium states, say denoted ω, which show condensation. In our case, this means that we have to show that the state has the following property: For some q $\rho_q = \omega(\alpha_q^* \alpha_q) > 0$, that is, the ω-expectation value of the number operator $a_q^* a_q$ of the q-mode particles is proportional to the volume. There are many possible techniques to reach this goal. We could apply correlation inequalities, we could use numerical methods, in due case all kinds of approximations, and so on. A particular way of proceeding is to solve the variational principle Eq. (3.1) or the energy-entropy criteria Eq. (3.7) for the equilibrium states. This is nothing new, so far. We can also ask the question whether it is possible to derive from these equilibrium criteria directly, or indirectly for that matter, a general independent equation for the crucial quantity ρ_q of our interest. Suppose that we can derive from general principles such an equation. Then we could try to solve this equation or try to show that this equation allows for a strictly positive valued solution for the condensate parameter ρ_q. If a positive answer can be given to this procedure, then we have shown the existence of Bose-Einstein condensation. Such an equation, if it can be derived, is called a *condensate equation*. In many cases solving or working with this equation is an economical way towards success in showing condensation.

We reproduce the proofs for the rigorous derivations of the q-mode condensate equation in its most general formulations [49, 162]. For more literature on this topic we refer also to [57, 31, 49] and references therein. By proceeding this way we indicate also that the notion of condensate equation and its position can be put in a larger context applicable to any other relevant order parameter for any boson model under consideration or for any other classical or quantum systems. We bring also into the picture the relation between the condensate equation and the time invariance of the state when dealing with ergodic states.

As explained, the condensate equation(s) should be derived for instance from the variational principle Eq. (3.1). It should be stressed that the derivation of the condensate equations is in fact an essential part of an explicit execution of the variation principle. It is simply the straightforward tool of performing explicitly the variation with respect to the variation of an order parameter of the model, in particular the condensate parameter. Here we have in mind for instance the parameter c linked to the condensate density for ergodic states given by $|c|^2 = \lim_V \omega(a_0^* a_0)/V = \rho_0$ in the case $q = 0$ Eq. (2.32).

Equally well, we should be able to derive these condensate equations from the energy-entropy criteria for equilibrium Eq. (3.7). That is what we have carried out in the derivations below. Nevertheless, for interpretational matters we make some direct connections with the variational principle. In any case the condensate equation for any homogeneous gauge invariant equilibrium state ω can be derived from the energy-entropy criterion. We specify the result for general homogeneous as well for the ergodic, extremal invariant, or the ergodic, symmetry-breaking, boson equilibrium states.

Preparing the formulation of the main result, for formal completeness let us start with a few technicalities, which should help us understand the underlying mathematical structure of the theory. First we remark that the following matrix elements for any choice of observables A, B, C,

$$\omega(A\delta(B)C) \doteq \lim_V \omega(A[H_V - \mu N_V, B]C) \tag{4.42}$$

should define operators $\delta(B)$ on a common domain $\mathscr{D}$ containing all the observables being polynomials in the creation and annihilation operators. This definition is mathematically understandable again in the sense of the GNS-representation ([26] and Eq. (7.1)) of the state ω. The linear map $\delta : B \to \delta(B)$ satisfies all the properties of a *derivation*, which are given by $\delta(AB) = \delta(A)B + A\delta(B)$ and by $\delta(A)^* = -\delta(A^*)$.

In the preceding subsection Eq. (4.12), we introduced the average operator $\alpha_q = \lim_V \alpha_{q,V}$ with $\alpha_{q,V} = \frac{a_q}{\sqrt{V}}$ where $\lim_V$ means the strong operator limit for the ω-state GNS-representation space $\mathscr{H}$. Remember that the existence of the operator is ensured for all states that are invariant under the (lattice) translation group, because the operator is a space average operator. We have in mind that limit Gibbs states are homogeneous and we learned that spontaneous symmetry breaking leads to ergodic or extremal translation (or lattice-)invariant states.

For convenience we repeat the following essential properties of the operator α_q Eq. (4.12) acting on the representation space $\mathscr{H}$ induced by the state ω:

1. The operator commutes with all creation and annihilation operators.
2. In particular, it is a normal operator: $\alpha_q^* \alpha_q = \alpha_q \alpha_q^*$.
3. Its relation to the condensate density operator is $n_q = \alpha_q^* \alpha_q$, where n_q is the q-density operator with q-condensate density in the state ω given by $\rho_q = \omega(n_q) > 0$.
4. It has a well-defined polar decomposition: $\alpha_q = U_q \sqrt{n_q}$ with U_q a unitary operator and $\sqrt{n_q} \geq 0$.
5. From its definition, it readily follows that it is not always translation invariant, but periodic in the q-direction: $\tau_y(\alpha_q) = e^{-iq \cdot y} \alpha_q$.

Now we are able to formulate a general property following directly from the EEB criterion of equilibrium.

Lemma 4.6. *Let ω be any homogeneous equilibrium state, satisfying the EEB-equilibrium condition Eq. (3.7). For any polynomial $P(a_k^*, a_k, \alpha_q^*, \alpha_q)$ in the creation and annihilation operators a_k and the α_q-operators, simply denoted by $P(a, \alpha)$, and for any polynomial $Q(\alpha_q^*, \alpha_q)$, simply denoted by $Q(\alpha_q)$, in the α_q-operators, we obtain*

$$\omega(P(a, \alpha)\delta(Q(\alpha_q))) = 0. \tag{4.43}$$

or explicitly written as

$$\lim_V \omega(P(a_k, \alpha_{q,V})[H_V - \mu N_V, Q(\alpha_{q,V})]) = 0$$

Proof. The proof is extremely simple. We consider an arbitrary complex number λ and substitute $X = P^* + \lambda Q$ in the equilibrium condition for the state ω Eq. (3.7). We then use the convexity of the ln-function to obtain the inequality $a \ln \frac{a}{b} \geq a - b$ for any pair (a, b) of positive real numbers and use the fact that the operators P and Q commute in order to obtain the inequality

$$\beta(|\lambda|^2 \omega(P\delta(P^*)) + \bar{\lambda}\omega(P\delta(Q)) + \lambda\omega(Q^*\delta(P^*)) + \omega(Q^*\delta(Q)))$$
$$\geq |\lambda|^2 \omega(PP^* - P^*P)$$

If we find that the unique λ-independent term $\omega(Q^*\delta(Q))$ in this inequality vanishes, then the theorem follows from this inequality by virtue of the vanishing of the terms linear in λ. We show as follows that this term indeed vanishes.

By taking $\lambda = 0$ in the inequality, we obtain $\omega(Q^*\delta(Q)) \geq 0$. After repeating the argument with Q replaced by Q^*, we also obtain $\omega(Q\delta(Q^*)) \geq 0$.

On the other hand, using the time invariance of the equilibrium state ω Eq. (3.8) and the fact that Q commutes with all other local operators, we obtain

$$0 = \omega(\delta(Q^*Q)) = \omega(\delta(Q^*)Q) + \omega(Q^*\delta(Q)) = \omega(Q\delta(Q^*)) + \omega(Q^*\delta(Q))$$

This equation together with the positivity of both terms yields $\omega(Q^*\delta(Q)) = 0$, finishing the proof of the Lemma.

Now we are in a position to obtain the condensate equation for the q-mode. In particular two versions of this equation are presented. The first one is intended for general limit Gibbs states, which are homogeneous states but not necessarily extremal space-invariant equilibrium states. A second version is derived for the extremal or ergodic equilibrium states, which are in due case non-homogeneous but q-periodic and which show spontaneous gauge symmetry breaking. For general limit Gibbs states we state the following:

Theorem 4.7. *Let ω be a general homogeneous limit Gibbs state satisfying Eq. (3.7). The q-condensate equation is given by*

$$\omega(\alpha_q^* \delta(\alpha_q)) = 0 \tag{4.44}$$

Proof. The formula of the theorem follows immediately from the preceding Lemma Eq. (4.43) by substituting $P = \alpha_q^*$ and $Q = \alpha_q$.

We should note the compact form of Eq. (4.44). We note also the entropy term of the variational principle, or of the energy-entropy criterion, does not contribute to the explicit form of the equation. The temperature enters only by way of the correlation functions. That Eq. (4.44) is a type of a condensate equation and should become clear shortly. In particular it should become clear after having been applied explicitly for some model Hamiltonians.

On the other hand, we could ask also for its relation to what is usually understood in the physics literature as the condensate equation. This equation is mostly referred to as the Euler equation resulting from the minimization of the free-energy density functional with respect to a variation in the condensate density variable (see [57] for example). This statement should be analyzed with some care. Nevertheless it remains instructive to link the result Eq. (4.44) directly to the general variational principle Eq. (3.1) of the free-energy functional with respect to a variation on the set of homogeneous states. In particular the identifying the type of variation of the states is interesting. Is the variation a change of the condensate density or is the variation related to another quantity? We now explain the relation between Eq. (4.44) and the general variational problem Eq. (3.1). As far as the latter is concerned, we again consider the free-energy density functional map f Eq. (3.1) defined on the state space: For any state σ

$$f : \sigma \to f(\sigma) = \lim_V \frac{1}{V} \{ \beta \, \sigma(H_V - \mu N_V) - S(\sigma_V) \}$$

with μ the chemical potential and $S(\sigma_V)$ the entropy of the restriction of the state σ to the finite volume V observables. The variational principle of statistical mechanics implies that each homogeneous (or periodic) equilibrium state ω_β minimizes the free energy density functional, or equivalently, for any arbitrary homogeneous state σ, $f(\omega_\beta) \leq f(\sigma)$. Next, consider the following inequality, which has been explicitly proven for lattice systems in [53], but immediately extendable to continuous systems and given by: Any homogeneous equilibrium state ω_β satisfies for any observable X the inequality

$$\lim_V \beta \, \omega_\beta(X^*[H_V - \mu N_V, X]) - \omega_\beta(X^*X) \ln \frac{\omega_\beta(X^*X)}{\omega_\beta(XX^*)}$$

$$\geq \lim_{\lambda \to 0^+} \frac{1}{\lambda} (f(\omega_\beta \circ \gamma_\lambda) - f(\omega_\beta)) \geq 0$$

where $\{\gamma_\lambda = e^{\lambda \Gamma} \mid \lambda \in \mathbb{R}^+\}$ is any one-parameter (namely λ) semigroup of homogeneous, completely positive maps (mapping states into states, see Eq. (7.2)) which map any ω_β-state (equilibrium state) into a perturbed one of the system, and where the map Γ on the set of observables is given by the action-type Lindblad generator Eq. (7.2)

$$\Gamma(.) = \lim_V \Gamma_V(.) = \lim_V \int_V dx ([\tau_x(X^*), .] \tau_x(X) + \tau_x(X^*)[., \tau_x(X)] \tag{4.45}$$

The notation τ_x stands again for the translation over the distance x. The proof of the inequality is a straightforward consequence of the biconvexity of the function $x, y \in \mathbb{R}^+ \to x \ln(x/y)$.

As a byproduct, this inequality shows explicitly in which sense the EEB criterion for the equilibrium state ω_β is related to the derivative at $\lambda = 0$ of the free-energy density functional defined on the set of homogeneous states.

Moreover we note that for the particular choice of $X = \alpha_q$ in Eq. (4.45), and using the result Eq. (4.44), the left hand side of Eq. (4.45) vanishes. This shows that Eq. (4.44) is equivalent to what could be called a quantum Euler equation of the variational principle for all state variations of this particular type. It is a classical analysis Euler equation in the parameter λ. The physical meaning and the effect of these variations can made clearer by the following direct computation, yielding $\gamma_\lambda(\alpha_q^\#) = \lim_V e^{\lambda \Gamma_V}(\alpha_{q,V}^\#) = e^{-\lambda}\alpha_q^\#$. The γ_λ-operation is nothing more than an example of a particular quantum dynamical semigroup (see Eq. (7.2)), mapping the homogeneous quantum states into themselves. A variation of the positive parameter λ makes the states $\omega_\beta \circ \gamma_\lambda$ wander around in the state space. In particular the map corresponds to the operation of multiplying by a real number $e^{-\lambda}$ the creation and the annihilation operators of a condensate particle, leaving all other observable operators $a_k^\#$ invariant (see also Eq. (7.2)). In particular, this means that the semigroup changes the value of the condensate density by a factor $e^{-2\lambda}$. We should stress the fact that this type of variation is a consequence of the special variation of creation and annihilation operators of the condensate mode only. In the case $q = 0$, this variation is therefore really related to a change of the parameter $c = \lim_V \omega(a_0/\sqrt{V})$ introduced and discussed before. Clearly all these arguments provide an alternative derivation of the existence of this particular type of Euler equation for the part of the variational principle leading to the condensate equation. This completes the discussion about the physical meaning of the result derived in Theorem Eq. (4.44) as well as about the question of whether Eq. (4.44) derived in the theorem coincides with the condensate equation as it was originally conceived. It has always been considered as being connected as an essential part of the variational principle. It is clear that we can apply this argument for the variations determining all one-point functions of the states. As mentioned, in the literature the original condensate equation referred only to the condensate variable ρ_0. It is clear from Eq. (4.43), Eq. (4.44) that we generalized the concept of the condensate equation to arbitrary q-condensation. We can deal with all cases $\rho_q \geq 0$, where $q = 0$ as well as $q \neq 0$.

Yet, below we give another non-trivial, nevertheless straightforward and immediate generalization of this variation procedure and hence of the condensate equation idea. Instead of taking only the operators α_q, we use now the space averages of any arbitrary local (or quasi-local) observable. In principle our methods lead to the derivation of possibly infinitely many new condensate equations. In practice, of course many of them lead us to trivialities or fail to give anything new. However for some boson models they represent essential and interesting contributions toward the solution of the boson problem under consideration. The least we can say that we can distinguish a condensate equation in a boson model each time there is another order parameter in the model. In the model studies below we find several applications of these condensate equations technique. Before proceeding to the applications, we discuss first the announced generalizations.

Take any local observable A and consider its space averages in the finite volume V, as well as in the limit $V \to \infty$:

$$\overline{A_V} = \frac{1}{V} \int_V dx\, \tau_x A \text{ and } \overline{A} = \lim_V \overline{A_V}$$

It is important to remark that these space averages $\overline{A}$ satisfy all the mathematical technical properties of the basic operators α_q used before in the proofs of Theorems

Eq. (4.43) and Eq. (4.44). First, they are also space averages. Furthermore, the operators $\overline{A}$ commute with all local observables because $[\overline{A}, a(f)] = 0$ for each local test function f in $\mathscr{S}$ and therefore they commute with all functions of the localized creation and annihilation operators. Finally the averages $\overline{A}$ satisfy the property of being normal operators, expressed by $[\overline{A}, \overline{A}^*] = 0$. For all these reasons we can repeat word by word the proof of Eq. (4.44) and get for arbitrary averages the following much stronger result yielding in principle an infinity of condensate equations.

Theorem 4.8. *Let ω be any homogeneous limit Gibbs state or equilibrium state, satisfying the EEB criterion for equilibrium at inverse temperature β, including the case $\beta = \infty$, which means that ω is a ground state. Let A be any local (or quasi-local) observable, then the A-condensate equation is given by*

$$\omega(\overline{A}^* \, \delta(\overline{A})) = 0$$

which reads more explicitly as follows

$$\lim_V \omega(\overline{A}_V^* \, [H_V(\mu), \overline{A}_V]) = 0 \tag{4.46}$$

This theorem yields the most general form of condensate equations. We should note that there is possibly an infinite number of such condensate equations, one for each local observable A.

To add to our understanding of the structural position of these condensate equations in the framework of solving the problem of equilibrium states, we give a new and full proof of the result of the Theorem Eq. (4.8), but now starting from a state satisfying the variational principle Eq. (4.45) instead of the energy-entropy criteria Eq. (3.7).

For each finite volume V, consider again the Lindblad generator (see Eq. (7.2))

$$\Gamma_V(.) = \int_V dx \{ [\tau_x \overline{A}_V^*, .] \tau_x \overline{A}_V + \tau_x \overline{A}_V^* [., \tau_x \overline{A}_V] \}$$

generating a dynamical semigroup of completely positive maps $\{ \gamma_{\lambda,V} = \exp \lambda \Gamma_V \, | \, \lambda \geq 0 \}$ mapping in an evident manner the set of locally normal states into itself. On the basis of the variational principle Eq. (4.45) we obtain for any locally normal state $\omega(.) = \lim_V tr \rho_V$

$$0 \leq \lim_{\lambda \to 0} \frac{1}{\lambda} (f(\lim_V \omega \circ e^{\lambda \Gamma_V}) - f(\omega))$$

$$\leq \lim_V \{ \beta \, tr \rho_V \overline{A}_V^* [H_V(\mu), \overline{A}_V] - tr \rho_V \overline{A}_V^* \overline{A}_V \ln \frac{tr \rho_V \overline{A}_V^* \overline{A}_V}{tr \rho_V \overline{A}_V \overline{A}_V^*} \}$$

The second inequality is again a consequence of the bi-convexity in the two variables of the function: $x, y \to x \ln(x/y)$.

Because of the normality of the operator $\overline{A}$, the second term of the right hand side of the inequality vanishes. This term represents the change of entropy density per unit of the parameter λ. The fact that this term vanishes means that the entropy term in the free energy does not play a role in the expression of the A-condensate equation. As already remarked before, the A-condensate equation does not depend on the entropy density, but only on the energy density. In turn, this means also that the entropy density does not depend explicitly on the A-condensate. This implies moreover that the derivation of the condensate equations and the condensate equations themselves holds as well for the ground $(T = 0)$-states as for the temperature $(T > 0)$-states. The condensate equations are, in this sense, equations which are independent from most of the other basic parameters of the system. In any case we get immediately the positivity property

$$0 \leq \lim_V \omega(\overline{A_V^*}[H_V(\mu), \overline{A_V}])$$

Analogously, we obtain the same inequality with $\overline{A_V}$ replaced by $\overline{A_V^*}$.

Using the same arguments as above but now working with the group of unitary operators $\{U_\lambda = \exp(i\lambda H_V(\mu))\}$, for all $\lambda \in \mathbb{R}$ and the corresponding dynamical group $\{\gamma_{\lambda,V} = \exp(i\lambda [H_V(\mu), .]) ; \lambda \in \mathbb{R}\}$ instead of with the semigroup, we also obtain from the variational principle the time invariance of the state (see also Eq. (3.8)) given by

$$0 = \lim_V \omega([H_V(\mu), X])$$

for each observable X of the system. In particular we find

$$0 = \lim_V \omega([H_V(\mu), \overline{A_V^* A_V}])$$
$$= \lim_V \{\omega_\beta([H_V(\mu), \overline{A_V^*}]\overline{A_V}) + \omega(\overline{A_V^*}[H_V(\mu), \overline{A_V}])\} \tag{4.47}$$

Using again the property that the averages commute with all local observables, we obtain the most general condensate equation announced in the theorem. This finishes the independent proof of Theorem Eq. (4.8) derived directly from the variational principle.

Finally we proceed to the special situation of the condensate equation only valid for extremal or ergodic equilibrium states. Let $\tilde{\omega}$ be any such ergodic state. Then it follows immediately from Theorem Eq. (4.8) that

$$0 = \tilde{\omega}(\overline{A^*})\tilde{\omega}(\delta(\overline{A})) \tag{4.48}$$

On the other hand we know already from Eq. (3.8) that

$$\tilde{\omega}(\delta(\overline{A})) = 0 \tag{4.49}$$

Hence for ergodic equilibrium states the condensate equation (4.8) gives nothing new or different from the time invariance property applied to the average operator $\overline{A_V}$. The connection between the condensate equation and the time invariance was already expected from the proof of the theorem Eq. (4.8). For these reasons we can call the time invariance property applied to the special choice of observables, namely the averages, the condensate equation for ergodic states.

Further on in this chapter we give a number of illustrations of the material about the condensate equations when we discuss explicitly a number of models like the free boson gas, the mean field boson gas, the Bogoliubov model, the model on super-radiance amplification by boson condensation, and others.

Before proceeding to these special boson models, it is interesting to mention that the ideas behind the derivation of these results, in particular in the form of the Theorem Eq. (4.8), about the most general form of the condensate equations theory developed so far, extend and are applicable in the search for equilibrium state solutions of all types of homogeneous quantum systems far beyond boson systems. We have in mind all kinds of quantum spin systems. In particular we think also about the applications for interacting fermion systems. In fact we presume that more interesting applications are expected in most models in which order parameters play a prominent role. These ideas about condensate equations are also extendable to homogeneous classical systems. This comes over to us as an wide open research domain in which these concepts and techniques have not been exploited so far.

4.4 Mean Field Bose Gas

The free Bose gas was the first boson model showing Bose-Einstein condensation. The model is important not only because it is the first model with this property. It is also fundamentally important because it shows the occurrence of this basic quantum phenomenon for a system of free bosons. It means that condensation is not an artifact of a special type of inter-particle interactions. Therefore the free Bose gas remains a statue of the physics literature, in particular for the field of quantum statistical mechanics.

On the other hand it is clear that the free Bose gas has its limitations as a physical theory, as well for the explanation of physical phenomena as because of its intrinsic theoretical properties. The theory behind the free Bose gas with its condensate was quickly considered the theory explaining the low temperature phenomenon of superfluidity. However the one-particle spectrum of the free Bose gas was quickly realized unsuitable for explaining superfluidity. (See Eq. (4.1) and e.g. [169] for a recent discussion about this topic.) From a more theoretical point of view the free Bose gas has also some less agreeable aspects making the model less useful even as a toy model. We mention the following reasons: Developing the microscopic theory of thermodynamics from the point of view of Gibbs states, we arrange matters directly or indirectly within the underlying philosophy that all three of the Gibbs ensembles (micro-canonical, canonical, grand-canonical) coincide in the thermodynamic limit. In this context the careful analysis by the Dublin group [101] (see also [170]) of the thermodynamic limit Gibbs states, keeping the grand canonical particle density ρ_{gr} constant, establishes a relationship between the grand canonical state ω_β^{gr} and the canonical states $\{\omega_\beta^\rho\}$ with densities ρ,

$$\omega_\beta^{gr}(.) = \int_0^\infty d\rho\, K(\rho, \rho_{gr})\, \omega_\beta^\rho(.) \tag{4.50}$$

where the function K is given by: Let ρ be any non-negative real number. Then

$$K(\rho,\rho_{gr}) = (\rho_{gr} - \rho_c)^{-1}\exp\{-\frac{\rho - \rho_c}{\rho_{gr} - \rho_c}\}, \text{ if } \rho > \rho_c$$

$$K(\rho,\rho_{gr}) = \delta(\rho - \rho_{gr}), \text{ if } \rho \leq \rho_c$$

Therefore if $\rho > \rho_c$ (the condensate region) the grand canonical equilibrium state ω_β^{gr} is a non-trivial convex combination of the canonical equilibrium states ω_β^ρ, where the canonical density ρ is the summation parameter.

We note that the condensate density is given by $\rho_0 = \rho_{gr} - \rho_c$ with ρ_c the critical density of the free Bose gas.

The function K is sometimes called the *Kac-function*. If the canonical density is larger than the critical density ($\rho > \rho_c$), then the Kac-function is not a trivial delta function, which means that in such a region the canonical and the grand canonical states are different states. This is a somewhat embarrassing situation for the free Bose gas in order to be a decent microscopic model for statistical mechanics yielding normal thermodynamical behavior. Moreover the Kac-function depends once more heavily on the boundary conditions. For attractive boundary conditions this Kac-function is computed in [160] and it turns out to be quite different from the corresponding Kac-density, which is computed in [101]. Nevertheless its main property of not being a trivial delta-function below the transition point remains in the attractive boundary conditions case.

Another annoying aspect of the free Bose gas model is related to the fact that the phase transition point (transition from no condensation to positive condensation) does not show a clear type of phase transition, whether first order or second order. It remains a point of discussion among theoreticians.

For all these reasons physicists were looking for other models which show Bose-Einstein condensation and which have preferably a decent thermodynamical behavior. Because the full two-body interaction model is unsolvable, a first solvable model showing BEC with good thermodynamical behavior is the mean field Bose gas. This model is also sometimes called the *imperfect boson gas*. For periodic boundary conditions the model has the following local Hamiltonian:

$$H_V^{mf}(\mu) = T_V - \mu N_V + \frac{\lambda}{2V}N_V^2$$

$$T_V = \sum_k \frac{k^2}{2m}a_k^* a_k \tag{4.51}$$

with $\lambda > 0$ the coupling constant. The grand canonical ensemble thermodynamic-limit state with constant density is computed in [54]. For this model the Kac-function is proven to be always a delta-function and therefore the canonical and grand canonical states coincide. The phase transition is a second-order transition. For all these reasons it is likely to consider the mean field model as the prototype microscopic statistical mechanics model for Bose-Einstein condensation. It is a solvable model for which we sketch how to find the solutions of the variational principle Eq. (3.1) as well as the solutions of the energy-entropy criterion for equilibrium Eq. (3.7). The solvability of the model makes that, in nearly all practical applications of Bose-

Einstein condensation models, the mean field Bose gas is ubiquitously present in the literature.

Let us start the analysis of the equilibrium states of this mean field model. First we note that the model Eq. (4.51) is indeed super-stable for all values of the chemical potential μ and hence for all total densities. Indeed for all constants $\mu_0 \leq 0$ we obtain the inequality

$$H_V^{mf}(\mu) = T_V - \mu_0 N_V + \frac{\lambda}{V}(N_V - \frac{(\mu - \mu_0)V}{2\lambda})^2; \; \lambda > 0$$

proving the stability of the mean field boson model.

We look first for the solutions of the variational principle Eq. (3.14). For each ergodic homogeneous state ω and for a given fixed grand canonical density $\rho = \lim_V \omega(N_V/V)$ we compute straightforwardly the value of the free-energy density functional for the state ω,

$$f(\omega) = \beta\{\int dk\, \varepsilon_k(r(k) - 1) + \frac{\lambda}{2}\rho^2\} - \int dk\, \{r(k) \ln r(k) - (r(k) - 1)\ln(r(k) - 1)\}$$

where $\varepsilon_k = \frac{k^2}{2m} - \mu$, $r(k) = \omega(a_k a_k^*)_t$ and $\rho = \rho_0 + \rho_c$, with $\rho_c = \int dk\,(r(k) - 1)$ and finally $\rho_0 = \lim_V \frac{\omega(a_0^* a_0)}{V} = \lim_V |\omega(a_0/\sqrt{V})|^2$ the condensate density. The energy density functional for any ergodic state depends clearly only on the one- and two-point functions of the state. Therefore the mean field model is a solvable model. By Eq. (3.14) the variational principle over all ergodic states is reduced to a variational principle over the set of quasi-free states $\mathfrak{Q}$. The variational parameters (see Eq. (3.14)) for this model are the continuous functions $r(k)$ satisfying $r(k) - 1 \geq 0$ and $\rho \geq \int dk\,(r(k) - 1)$. The particle density ρ is a priori given and fixed. Remains the variations over the set of one-point functions which consists of the condensate parameter $\sqrt{\rho_0}$ and the phase factor of the one-point function.

Note that the free energy density of the mean field model does not depend on this phase, which therefore can take any arbitrary real value.

Next look for the variations with respect to the condensate parameter $|c| = \sqrt{\rho_0}$. As explained before, this variation is performed by considering the corresponding condensate equation (4.44). Since we have a closed boson system, we only expect $(k = 0)$-condensation. As in the free boson gas, we prove that there is indeed no $(k \neq 0)$-condensation. Indeed for each $k \neq 0$m, we get for all translations τ_x: $\omega(a_k) = \omega(\tau_x a_k) = e^{ikx}\omega(a_k)$. Hence $\omega(a_k) = 0$ on the basis of the translation invariance of the state. None of the $(k \neq 0)$-modes shows condensation.

A direct computation yields the zero-mode condensate equation, that is,

$$\rho_0(\mu - \lambda\rho) = 0$$

The density ρ being fixed this equation has two different solutions, namely (i) $\rho_0 = 0$ and (ii) $\rho_0 > 0$ yielding in the second case the explicit relation between the chemical potential and the given total particle density: $\mu = \lambda\rho$.

The first case corresponds to an absence of boson condensation, so that the free energy density is completely described solely by the function $r(k)$.

The second case corresponds to the condensate region, with $\mu = \lambda\rho$. This equality is a basic relation if there is condensation. In essence it reduces the variational problem of this model to that of the free boson gas. We note however the essential difference in the condensate regions between the free Bose gas, where the chemical potential $\mu = 0$ holds, and the mean field boson gas, where we have a strictly positive chemical potential $\mu = \lambda\rho > 0$. This fact is at the origin of the fact for the mean field model, the Gibbs ensembles coincide, where they do not coincide for the free boson gas.

Now we are prepared to perform the variations with respect to the variables $r(p)$. We obtain straightforwardly the corresponding Euler equation

$$\beta(\frac{p^2}{2m} - \mu + \lambda\rho) = \ln\frac{r(p)}{r(p) - 1}$$

an equation which we can write in the more recognizable form

$$r(p) - 1 = \frac{1}{e^{\beta(\frac{p^2}{2m} - \mu + \lambda\rho)} - 1}$$

We use the double commutator inequality Eq. (3.10) and we derive for all $p \neq 0$:

$$E_p \equiv \lim_V \omega([a_p, [H_V, a_p^*]]) = \frac{p^2}{2m} - \mu + \lambda\rho \geq 0$$

The E_p are the energy spectral values of the boson quasi-particles (see also Eq. (5.1)) of the mean field model for all non-condensed modes. We note that in the condensate region ($\rho_0 > 0$), the spectrum coincides with that of the free boson gas, namely $E_p = \varepsilon_p$, yielding also the same critical density ρ_c. Outside the condensate region ($\rho_0 = 0$), we obtain the following explicit relation between the given density ρ and the chemical potential μ:

$$\rho = \int dp\,(r(p) - 1) = \int dp\,\frac{1}{e^{\beta(\frac{p^2}{2m} - \mu + \lambda\rho)} - 1}$$

All this solves completely the variational principle of statistical mechanics for the mean field model.

In order to write explicitly the ergodic or extremal space invariant, but gauge-breaking, states we use Eq. (4.19) and keep in mind Theorem Eq. (4.2). We note that the condensation occurs only in the ($q = 0$)-mode, as for the free boson gas. This expresses also that the space translation invariance is not spontaneously broken.

The result of the variational principle is a solution given by the quasi-free states $\{\omega_\alpha | \alpha \in [0, 2\pi]\}$ with non-trivial two-point functions, determined as the functions $r(p)$ which is found above, and with $s(p) = 0$. Using Eq. (2.31), the quasi-free state has the one-point function given by $\omega_\alpha(a(f)) = \widehat{f}(0)\sqrt{\rho_0}\,e^{i\alpha}$.

For didactic reasons let us also look for the solutions of the energy-entropy balance criterion for equilibrium Eq. (3.7) for our mean field model.

Let us consider again any ergodic state ω. First we use the condensate equation (4.44) for $\alpha_{q=0}$, yielding again $\rho_0(\mu - \lambda\rho) = 0$. This means that we once again produce the two possibilities: $\rho_0 = 0$ (no condensation), or $\rho_0 > 0$ and $\mu = \lambda\rho$. Both cases compel us to look for the other modes $p \neq 0$. Therefore we substitute $X = a_p$ in Eq. (3.7), leading straightforwardly to the inequality

$$-\beta E_p \geq \ln \frac{r(p) - 1}{r(p)}$$

We next substitute $X = a_p^*$ in Eq. (3.7), and compute again. We obtain the analogous inequality but with the opposite inequality sign. This means that we obtain an equality and that $r(p)$ is the same function as found above in the case of the variational principle. This is an illustration of the fact that both equilibrium criteria Eq. (3.1) or Eq. (3.14) and Eq. (3.7) yield exactly the same result for the homogeneous mean field model.

Let us conclude the study of the mean field boson model by writing down the self-consistency equation relation for the total density ρ of the equilibrium state ω. By means of the obtained results it becomes

$$\rho = \lim_V \omega(\frac{N_V}{V}) = \rho_0 + \int dp \, \frac{1}{e^{\beta E_p} - 1}$$

where $E_p = \frac{p^2}{2m} - \mu + \lambda\rho$. Hence we arrive at an expression equivalent to that for the free Bose gas, but where the one-particle energies ε_p are replaced by the quasi-particle energies $\varepsilon_p + \lambda\rho$. For more details about spectrum considerations see Chapter Eq. (5). Our analysis has uncovered two regimes:

(i) The condensate phase region with $\rho_0 > 0$ and $\mu = \lambda\rho$. The total density equation becomes

$$\rho = \rho_0 + \int dp \, \frac{1}{e^{\beta \frac{p^2}{2m}} - 1} = \rho_0 + \rho_c$$

where ρ_c is the critical density of the free Bose gas. The density equation is the same as for the free boson gas.

Hence also in the condensate region of the mean field model we get $\lim_{p \to 0} \frac{E_p}{p} = 0$; the Landau criterion for superfluidity Eq. (4.1) is not satisfied.

(ii) The normal phase region with $\rho_0 = 0$. Then $\lambda\rho - \mu \geq 0$ and the density equation becomes

$$\rho = \int dp \, \frac{1}{e^{\beta(\frac{p^2}{2m} - \mu + \lambda\rho)} - 1}$$

which is the density equation of a free gas of quasi-particles. The latter are the same particles as the free boson gas particles but carrying a modified one-particle energy E_p.

4.5 Super-radiance and Matter Wave Amplification

In this section we discuss a solvable model that explains the discovery of the effects of Dicke super-radiance on (a) boson condensates and (b) the phenomenon of BEC matter wave amplification (see [88] and references therein). In these experiments the condensate atoms are illuminated by a laser beam. The condensed atoms scatter the photons coming from the beam and receive their corresponding recoil momentum on the basis of momentum conservation. For more details about the experiments and their physical phenomenological theories see also the references [137, 139, 140].

Apart from the physics covered by the model, which we present, we consider the model here as a solvable model which shows condensation in a $(q \neq 0)$-mode. As a consequence of Theorem Eq. (4.5) we get spontaneous space translation symmetry breaking in the presence of condensation. Afterwards we discuss the gauge symmetry breaking. The model is given by the following local Hamiltonian written down in one space dimension $(d = 1)$ and with periodic boundary conditions:

$$H_V^{sr}(\mu) = \sum_k (\varepsilon_k - \mu) a_k^* a_k + \Omega \, b_q^* b_q + \frac{\lambda}{2V} N_V^2 - \frac{g}{\sqrt{V}} (a_q^* a_0 b_q + a_q a_0^* b_q^*) \quad (4.52)$$

The first term is the kinetic energy term of the bosons with one-particle energies $\varepsilon_k = \frac{k^2}{2m}$; the second term is the kinetic energy of a single mode photon laser with frequency Ω; the third term is a system-stabilizing boson mean field term. The last term describes the interaction between the laser mode and the boson $q = 0$-mode and the $q \neq 0$-mode. The λ and g are positive coupling constants.

The total local Hamiltonian Eq. (4.52) is clearly space homogeneous and there exists a global gauge symmetry for the bosons. Considering the gauge transformations of the 0-mode and the q-mode independently, we note that both of them are broken already in the Hamiltonian.

We are interested in the equilibrium states of the total system. In particular, we look for the ergodic or extremal equilibrium states for the bosons as well as for the laser mode which we consider as a collective or macroscopic boson mode. These ergodic states are chosen to be product states with respect to the product system nature of the boson system and the photon system.

The operator $a_q^* b_q / V$ and the operator $a_0 / \sqrt{V}$ are space averages of local operators and therefore the value of the energy density for any ergodic state ω equals

$$e(\omega) = \int dk \, \varepsilon_k \omega(a_k^* a_k) + \Omega \lim_V \omega(\frac{b_q^* b_q}{V}) + \frac{\lambda}{2} \rho^2 + g \{ \omega(\frac{a_0}{\sqrt{V}}) \lim_V \omega(\frac{a_q^* b_q}{V}) + h.c. \}$$

where ρ denotes again the total density of the bosons. We note that the energy density formula is given in terms of the one- and two-point functions of the state and that therefore the system is a solvable model.

If there is condensation in the q-mode then the space translation symmetry is spontaneously broken down to a lattice symmetry in the q-direction Eq. (4.5), with a period $\gamma = \frac{2\pi}{|q|}$. The Hamiltonian Eq. (4.52) is gauge breaking for the single 0-mode gauge transformations of the bosons. This mode on the other hand remains itself space homogeneous. For the boson system we obtain periodic states with period γ; the same holds for the laser mode. Using any such ergodic state ω, we compute the limits

$$\lim_V \omega(\frac{a_q^*}{\sqrt{V}}) = \frac{1}{V_\gamma}\int_{V_\gamma} dx\, e^{iq.x}\omega(a^*(x))$$

with the integral over the cubic box lattice cell with sides equal to γ and with volume V_γ. Using the expression Eq. (2.31) of the one-point function, we find $\sqrt{\rho_q}e^{i\varphi_q} = \lim_V \omega(\frac{a_q^*}{\sqrt{V}})$. The previous result holds analogously for the two other densities. We introduce their corresponding densities ρ_0 attached to the mode operator a_0 and $\widetilde{\rho}_q$ to the density(intensity) related to the laser operators $b_q^\#$.

In what follows, we set all phases equal to zero to simplify notation. This corresponds with performing the right gauge transformations for the three corresponding modes. In fact the phases should satisfy the relation $\varphi_q - \varphi_0 - \widetilde{\varphi}_q = 0$ to reach the minimal energy state.

Consider first the condensate equations Eq. (4.44), or Eq. (4.8) for the three mode operators a_0^*, a_q^* and b_q^*, appearing in the interaction. We get, respectively,

$$(-\mu + \varepsilon_0 + \lambda\rho)\rho_0 = g\sqrt{\rho_0\rho_q\widetilde{\rho}_q}$$

$$(-\mu + \varepsilon_q + \lambda\rho)\rho_q = g\sqrt{\rho_0\rho_q\widetilde{\rho}_q}$$

$$\sqrt{\widetilde{\rho}_q}(\sqrt{\widetilde{\rho}_q} - \frac{g}{\Omega}\sqrt{\rho_0\rho_q}) = 0$$

Of course, we denote by $\widetilde{\rho}_q = \lim_V \widetilde{\omega}(\frac{b_q^*b_q}{V})$ the intensity, also called the q-mode laser density or the density of the photons, and denote again by ρ the total density of the boson particles. We are interested in solutions of these equations as a function of the chemical potential μ, which itself is fixed by the density of bosons constraint expressed by the constant density formula $\rho = \lim_V \omega(\frac{N_V}{V})$ with $N_V = \sum_k a_k^* a_k$.

The following small-scale analysis of the equations teaches us the following: Suppose that one of the densities ρ_0, ρ_q, or $\widetilde{\rho}_q$ vanishes. Then the situation is uninteresting because the interaction term disappears in the energy density formula and the variational principle yields a system of free bosons about which we already know everything. In that case only condensation in the 0-mode is possible if the total boson density is large enough and if the dimension is larger than 2. If the dimension is smaller than 2, there is no condensation (see Eq. (4.1)).

Next we look for the possibility of a non-trivial solution. Considering the model Hamiltonian, the boson system is a subsystem of the total system consisting of the boson system and the laser system. If we restrict our attention to the boson system, we find that it is an open system in contact with the laser system. In order that the photon system is thermodynamically in balance with the boson system, or in other words that it is effective, it should be a one-mode macroscopic system. Therefore it is natural to assume that this mode has a finite density $\widetilde{\rho}_q \neq 0$. This is assumed as a starting point of our analysis and we do not dig further into the details of the mechanisms of the laser system. This situation should be considered as an initial condition expressing the physical property that the laser radiation is sufficiently intensive.

The third condensate equation yields $\rho_0\rho_q = \widetilde{\rho}_q\Omega^2/g^2$, which also implies that the other densities are non-vanishing, that is, $\rho_0 \neq 0$ and $\rho_q \neq 0$. Now using the second condensate equation,

$$\rho_0 = \frac{\Omega}{g^2}(\varepsilon_q + \lambda\rho - \mu)$$

which expresses that the condensate density ρ_0 depends linearly on the boson excitation energy ε_q. Substituting this result into the second equation again,

$$\rho_q = \frac{\Omega}{\varepsilon_q + \lambda\rho - \mu}\tilde{\rho}_q$$

which expresses a direct proportionality of the q-condensate ρ_q with the laser density $\tilde{\rho}_q$.

Let us now consider the non-condensate modes $k \neq 0$ or $k \neq q$. We should immediately recognize that we obtain for this part of the system a pure mean field type model analogous to the one treated in detail in the preceding section. Using as above the equilibrium conditions, for example using the energy-entropy balance conditions Eq. (3.7) and the double commutator inequality Eq. (3.10), in which we substitute the observable X by a_k^*, we derive immediately that, for all k, $\varepsilon_k + \lambda\rho - \mu \geq 0$. In particular, $\lambda\rho - \mu \geq 0$ holds. This shows already that the above expressions for ρ_0 and ρ_q are meaningful solutions. In fact we can say more. From the first condensate equation together with the third condensate equation we obtain the exact and explicit relation between the total boson density ρ and the chemical potential μ: Using $\varepsilon_0 = 0$,

$$\lambda\rho - \mu = \frac{1}{2}\{\sqrt{\varepsilon_q^2 + 4g^2\tilde{\rho}_q} - \varepsilon_q\}$$

Note that this equation yields $\lambda\rho - \mu > 0$. We should compare this result with the corresponding one obtained for the pure mean field model. Only in the case that $g = 0$ and/or $\tilde{\rho}_q = 0$ do we obtain the mean field result $\lambda\rho - \mu = 0$.

Proceeding further with the non-condensate modes, as we did above for the pure mean field model, we derive the same function r:

$$r(k) - 1 = \frac{1}{e^{\beta(\varepsilon_k + \lambda\rho - \mu)} - 1}$$

Exactly because $\lambda\rho - \mu > 0$, the critical density is finite in all the dimensions $d \geq 1$ and the model represents a meaningful theory in all dimensions.

In any case, the ergodic equilibrium states, denoted by ω, of this boson system are now completely known. They are quasi-free states of the boson system determined by the truncated two-point function

$$\omega(a(f)a^*(f))_t = \int dk\, \overline{\hat{f}}(k)\hat{g}(k)r(k)$$

and by the one-point function with the reintroduced phases φ_0 and φ_q, given by

$$\omega(a^*(f)) = \widehat{f(0)}\sqrt{\rho_0}e^{i\varphi_0} + \widehat{f(q)}\sqrt{\rho_q}e^{i\varphi_q}$$

The quantities ρ_0, ρ_q, $\lambda\rho - \mu$ and the function r are determined as a function of the photon laser density $\tilde{\rho}_q$ and the total boson density ρ.

This model has been introduced in the paper [140], where the analysis of the model is performed without using explicitly the condensate equations contrary to what is done here. Comparing the two methods illustrates the power and the elegance of working with the method of the condensate equations for the study of condensation in boson models in general. Nevertheless, those of us interested in a complete analysis of the model thermodynamics in terms of the total density ρ, as well as for more information about the physical phenomena that lie at the origin of the conception of this model, should consult the reference [140].

4.6 Theory of Bogoliubov

The leading motivation of Bogoliubov to introduce his theory was to explain the phenomenon of superfluity while keeping in mind that the basic ingredient was boson condensation and that the free Bose gas condensation state, following Landau's criterion Eq. (4.1), failed to yield the appropriate spectrum explaining superfluidity. Therefore Bogoliubov [20, 21, 22, 23] tried to conceive an interacting boson model Eq. (2.4), which nevertheless was solvable. He started from the solvable free gas showing condensation and switched on an interaction between the boson particles. This interaction should not have destroyed the condensate. To the contrary, the latter should have played the dominant role in the construction of the effective interactions between all particles.

In other words only interactions intermediated by condensate particles are considered relevant and all direct particle interactions between excited particles can be disregarded.

Bogoliubov's model has had an enormous impact on the activities in quantum many-body boson theory in general and on the understanding of superfluidity in particular. For a long time it remained the only reliable and workable theory available, although it had too many theoretical diseases [8, 169, 9] to be considered a modern theory of a quantum dynamical system. In particular, in [8] it is observed that the original Bogoliubov model was not a thermodynamically stable model. Adaptation to ensure the property of super-stability, leads to the following Bogoliubov model Hamiltonian with periodic boundary conditions:

$$H_V^{Bog} = \sum_k \varepsilon_k a_k^* a_k + \frac{1}{2V} \sum_{k \neq 0} v(k)(a_k^* a_{-k}^* a_0 a_0 + a_0^* a_0^* a_k a_{-k})$$

$$+ \frac{1}{V} a_0^* a_0 \sum_{k \neq 0} v(k) a_k^* a_k + \frac{1}{2V} \sum_{k,k'} a_k^* a_{k'}^* a_{k'} a_k \qquad (4.53)$$

where $v(k)$ is again the Fourier transform of the two-body potential (see Eq. (2.4)), which is chosen integrable satisfying the properties $0 \leq v(k) = v(-k) \leq v(o)$, for k inside, and $v(k) = 0$ outside a bounded region in k-space. The second and the third term in the Hamiltonian are the genuine Bogoliubov model terms. They contain only interactions between the ground mode $q = 0$ and the excited modes $k \neq 0$. There is no direct interaction between the excited modes. The last term in the interaction is recognized to be essentially equal to $N_V(N_V - 1)/V$, which is manifestly of the mean field interaction type treated above, and which guarantees the stability of the refreshed Bogoliubov model Eq. (4.53). This term is added to the original Bogoliubov model Hamiltonian. The idea of stabilizing the Bogoliubov theory was already present in older works [59, 171, 110] and was explicitly discussed in [8]. In all the older works [110], the analysis of the model with this stabilizing term has never been complete. It is interesting to remark that the super-stability of the model Eq. (4.53) is a direct consequence of the following simple operator equality: For $k \neq 0$,

$$a_0^* a_0 (a_k^* a_k + a_{-k}^* a_{-k}) + a_k^* a_{-k}^* a_0 a_0 + a_0^* a_0^* a_{-k} a_k$$
$$= (a_0^* a_k + a_{-k}^* a_0)^* (a_0^* a_k + a_{-k}^* a_0) - a_k^* a_k - a_0^* a_0$$

yielding straightforwardly the inequality

$$a_k^* a_{-k} a_0 a_0 + a_0^* a_0^* a_{-k} a_k + a_0^* a_0 (a_k^* a_k + a_{-k}^* a_{-k}) \geq -a_k^* a_k - a_0^* a_0$$

The model Hamiltonian Eq. (4.53) is clearly space translation and gauge invariant for the full gauge group $U(1)$. Again we must look for the equilibrium states of the model. Therefore we look for the extremal space invariant or ergodic equilibrium states satisfying one of the two or both equilibrium conditions Eq. (3.1), Eq. (3.7) in the thermodynamic limit which we took keeping the density of particles $\rho = \omega(N_V/V)$, for all volumes V, constant as well as in the limit $V \to \infty$.

First we compute the energy density $e(\omega)$ for any ergodic state ω. It takes the form

$$e(\omega) - \mu\rho = \int dk \, (\varepsilon_k - \mu) \omega(a_k^* a_k)$$
$$+ \frac{1}{2} \int dk \, v(k) (\omega(a_k^* a_{-k}^*) \rho_0 e^{2i\varphi(0)} + h.c.)$$
$$+ \rho_0 \int dk \, v(k) \omega(a_k^* a_k) + \frac{1}{2} v(0)\rho^2 \qquad (4.54)$$

where $\varphi(0) = \arg(\lim_V \omega(a_0^*/\sqrt{V}))$. We denote in a similar way by $2\varphi(k)$ the phases of the functions $\omega(a_k^* a_{-k}^*)$.

As noted, this energy density depends only on the one- and two-point functions of the state and therefore the model is solvable. The general variational principle for equilibrium is reduced to the principle acting on the set of quasi-free states Eq. (3.14). Arguing on the basis of the energy-entropy criterion Eq. (3.7) for equilibrium yields its corresponding simplifications.

We now continue with the analysis of the variational principle criterion. It is good to keep in mind that this variational principle is a principle of minimality of the free energy functional, which depends on the continuous functions $r(k) \geq 1$ with constraint $\int dk\,(r(k) - 1) \leq \rho$, the continuous functions $t(k)$ satisfying

$$0 \leq t(k) = \sqrt{(r(k) - 1)r(k) - |s(k)|^2} \leq \sqrt{(r(k) - 1)r(k)}$$

and finally we consider minimality variation with respect to the order parameter $\sqrt{\rho_0}$ and the phases.

Minimization of the functional with respect to the phases implies immediately that the minimal energy is reached for the following choice of the phases: $\varphi(k) + \varphi(-k) - 2\varphi(0) = \pi$ for all k. At the end of the computations we can reintroduce more general phases by means of suitable gauge transformations always satisfying this relation. Having done that, we obtain for the pairing order parameter $\lim_V \omega(a_k a_{-k}) = -\lim_V |\omega(a_k a_{-k})| = s(k)$ and for the condensate order parameter $\lim_V \omega(a_0/\sqrt{V}) = \sqrt{\rho_0}$, which are both real numbers.

Before going on, we mention that the variational principle has already been used to solve an approximated version of the above model Eq. (4.53) in the paper [7]. The approximation consists of making the famous Bogoliubov approximation in the genuine Bogoliubov terms. It consists of replacing the annihilation operator a_0 by a complex number. If we take into account the choice of the phases, we get that the expectation value $\omega(a_0) = c\sqrt{V}$ where $c = \sqrt{\rho_0}$ is a non-negative real number. At the end of the analysis of this model we come back to this point and discuss the important implications that are a consequence of this approximation.

We proceed with the analysis of the model Eq. (4.53). Exactly as we did for the previous models, we try to extract information about the system by considering possible condensate equations Eq. (4.8). After inspection of the model Hamiltonian, three probable relevant space averages manifest for this procedure.

There is of course the 0-mode average

$$\frac{a_0}{\sqrt{V}} = \frac{1}{V} \int_V dx\, a(x)$$

There is also the pairing average: Let $Q_V \equiv \sum_{k \neq 0} v(k) a_k a_{-k}$ and check that

$$\lim_V \frac{Q_V}{V} = \lim_V \frac{1}{V} \int_V dy\, \tau_y(a(v)a(x=0))$$

Hence the operator Q_V/V behaves as a space average in the limit V tending to infinity. Finally denote $P_V \equiv \sum_k v(k) a_k^* a_k$. Then

$$\lim_V \frac{P_V}{V} = \lim_V \frac{1}{V} \int_V dy\, \tau_y(a^*(v)a(x=0))$$

also behaves as an average.

A straightforward computation yields that the condensate equation related to the average P_V/V is trivial. We can understand this result because P_V is linked to the local generator of the space translations forming a symmetry group of the model. Therefore the number of relevant condensate equations reduces to two equations.

Computing the condensate equation for $a_0/\sqrt{V}$, we note that the resulting equation is equivalent to the stationarity Equation (3.8) for this observable, which is given by

$$\{-\mu + v(0)\rho + \int dk\, v(k) s(k) + \int dk\, v(k)(r(k) - 1)\}\rho_0 = 0 \qquad (4.55)$$

Analogously, the stationarity of the state for the observable Q_V/V is given by

$$\int dk\,(\varepsilon_k - \mu + v(0)\rho)v(k)s(k) + \rho_0 \int dk\,v(k)^2(s(k) + r(k) - \frac{1}{2}) = 0 \qquad (4.56)$$

implying its corresponding condensate equation.

The double commutator inequality Eq. (3.10) yields the following information: For the zero mode $k = 0$,

$$\lim_V \omega([a_0, [H_V^{Bog}(\mu), a_0^*]]) = -\mu + v(0)\rho + v(0)\rho_0 + \int dk\,v(k)(r(k) - 1) \geq 0 \quad (4.57)$$

and for $k \neq 0$,

$$\lim_V \omega([a_k, [H_V^{Bog}(\mu), a_k^*]]) = \varepsilon_k - \mu + v(0)\rho + v(k)\rho_0 \geq 0 \qquad (4.58)$$

The first non-trivial conclusions from this information are the following: Eq. (4.55) leads to distinguishing the two cases: (i) $\rho_0 = 0$ and (ii) $\rho_0 > 0$.

Let us first consider the case of absence of condensation: $\rho_0 = 0$. From Eq. (4.56), $\int dk(\varepsilon_k - \mu + v(0)\rho)v(k)s(k) = 0$. From Eq. (4.58), it follows readily that the function $s(k) = 0$ almost everywhere. We therefore conclude that $\rho_0 = 0$ implies $s(k) = 0$, or the absence of boson condensation implies the absence of the boson pairing phenomenon. In this case the model reduces to the mean field model.

We should realize that $\omega(a_k a_{-k}) \neq 0$ corresponds to boson pairing, that is, the boson counterpart of the better known fermion pairing in the Bardeen-Cooper-Schriefer(BCS)-theory.

We next consider the case of condensation, $\rho_0 > 0$. Suppose on the other hand that the boson-pairing order parameter vanishes as well, or that $s(k) = 0$ almost everywhere. Then it follows from Eq. (4.56) that

$$\rho_0 \int dk\,v(k)^2(r(k) - \frac{1}{2}) = 0$$

But that is in contradiction with the positivity of the state expressed by $r(k) - \frac{1}{2} \geq \frac{1}{2}$. Therefore $s(k) \neq 0$. Combing both arguments, we conclude that there is a non-trivial condensation if and only if there is a non-trivial pairing. In other words, we can conclude that the theory of Bogoliubov shows always a double phenomenon of condensations, namely boson condensation and pairing condensation, or none of them.

After this intermediate result, we continue the search for the ergodic equilibrium states ω satisfying the variational principle Eq. (3.14). As for any translation τ_x holds $\omega(\tau_x a_k) = e^{ikx}\omega(a_k) = \omega(a_k)$ we have that $\omega(a_k) = 0$ for all $k \neq 0$. Hence the equilibrium states do not break the gauge symmetry for the single $(k \neq 0)$-modes, consistent with absence of condensation in these modes.

As $\varepsilon_{-k} = \varepsilon_k$ and $v(-k) = v(k)$, we can let $\omega(a_k^* a_k) = \omega(a_{-k}^* a_{-k})$. Considering the formula Eq. (4.54), it follows from Eq. (3.14) that the equilibrium states are quasi-free states determined by the one- and two-point functions, for $k = 0$, given by $\omega(a_0), \omega(a_0^* a_0), \omega(a_0 a_0)$, and for $k \neq 0$, by the functions $\omega(a_k^* a_k), \omega(a_k a_{-k})$.

Following the idea of the proof of Eq. (2.6), consider a canonical Bogoliubov transformation Eq. (7.3) of the form

$$a_k = \tilde{a}_k \cosh \alpha_k + \tilde{a}^*_{-k} \sinh \alpha_k$$

transforming the boson variables a_k into the new boson variables $\tilde{a}_k$. The real functions $\alpha(k)$ are chosen to satisfy $\alpha_k = \alpha_{-k}$, and we look for such a transformation function which should satisfy the property that $\tilde{s}(k) \equiv \omega(\tilde{a}_k \tilde{a}_{-k}) = 0$. Later we fix the functions $\alpha(k)$ completely. We compute explicitly

$$\omega(a^*_k a_k) = (\tilde{r}(k) - \frac{1}{2}) \cosh 2\alpha_k - \frac{1}{2}$$

$$\omega(a_k a_{-k}) = (\tilde{r}(k) - \frac{1}{2}) \sinh 2\alpha_k$$

where, conforming with previous notation, $\tilde{r}(k) \equiv \omega(\tilde{a}_k \tilde{a}^*_k)$. The free energy density of the model becomes

$$f(\omega) = \int dk \, (\varepsilon_k + \rho_v(k))\{(\tilde{r}(k) - \frac{1}{2}) \cosh 2\alpha_k - \frac{1}{2}\}$$

$$- \rho_0 \int dk \, v(k)\{(\tilde{r}(k) - \frac{1}{2}) \sinh 2\alpha_k\} - \mu\rho + \frac{1}{2}v(0)\rho^2 \tag{4.59}$$

$$- \frac{1}{\beta} \int dk \, \{\tilde{r}(k) \ln \tilde{r}(k) - (\tilde{r}(k) - 1) \ln(\tilde{r}(k) - 1)\} \tag{4.60}$$

and the constraint of constant density, equal to ρ, becomes

$$\rho = \rho_0 + \int dk \, \{(\tilde{r}(k) - \frac{1}{2}) \cosh 2\alpha_k - \frac{1}{2}\}$$

Now we look for the extremum equation (the quantum Euler equation) for the variations with respect to the set of functions $\{\tilde{r}(k) \geq 1\}$ satisfying the density constraint. It is obtained from Eq. (4.59) by a standard variational computation yielding

$$\frac{\tilde{r}(k)}{\tilde{r}(k) - 1} = \exp \beta \{(\varepsilon_k + \rho_0 v(k) - \mu + v(0)\rho) \cosh 2\alpha_k - \rho_0 v(k) \sinh 2\alpha_k\}$$

We denote $x = v(0)\rho - \mu$, $f_k = \varepsilon_k + x + \rho_0 v(k)$, $h_k = \rho_0 v(k)$ and note that the double commutator inequality Eq. (3.10) yields $f_k \geq 0$ for all k. In particular, we find that $x + \rho_0 v(k) \geq 0$ as well as $h_k \geq 0$. Next we fix the function α_k according to the relation $\coth 2\alpha_k = \frac{f_k}{h_k}$ or equivalently by

$$\cosh 2\alpha_k = \frac{f_k}{E_k} \; ; \; \sinh 2\alpha_k = \frac{h_k}{E_k}$$

with the notation

$$E_k = \sqrt{f_k^2 - h_k^2} = [(\varepsilon_k + x)(\varepsilon_k + x + 2\rho_0 v(k))]^{\frac{1}{2}} \geq 0 \tag{4.61}$$

The function $\tilde{r}(k)$ becomes

$$\tilde{r}(k) = \frac{e^{\beta E_k}}{e^{\beta E_k} + 1} \tag{4.62}$$

and the constant density equation becomes

$$\rho = \rho_0 + \frac{1}{2} \int dk \left[\frac{\varepsilon_k + x + \rho_0 v(k)}{E_k} \coth \frac{\beta E_k}{2} - 1 \right] \tag{4.63}$$

These formulae indicate that the values of the E_k constitute the spectrum of the harmonic collective excitations or of the quasi-particles (see Eq. (5)) of the Bogoliubov model. We refer to the next chapter for more elaborate discussions about the dynamics and spectrum of solvable models of the type like the Bogoliubov model. All values of this spectrum should be real and non-negative for all k. This follows by applying the general and by now known double commutator inequality Eq. (3.10): $\lim_V \omega([X^*, [H_V^{Bog} - \mu N_V, X]]) \geq 0$ holding for each equilibrium state and for each observable X. Next, we let $X = \tilde{a}_k$, where $\tilde{a}_k = z_k a_k - w_k a_{-k}^*$, and z_k and w_k are arbitrary complex numbers satisfying $|z_k|^2 - |w_k|^2 = 1$. Then

$$\lim_V \omega([X^*, [H_V^{Bog} - \mu N_V, X]]) \geq 0 \text{ iff } f_k^2 - h_k^2 \geq 0$$

The density constraint equation (4.63), together with the condensate equations Eq. (4.55) and Eq. (4.56), determine the condensate density parameter ρ_0 as well as the pairing function $s(k)$. For simplicity we continue the analysis for dimension $d = 3$. Consider the energy spectral function $E_k = [(\varepsilon_k + x + \rho_0 v(k))^2 - \rho_0^2 v(k)^2]^{1/2}$. It depends on the two parameters ρ_0 and x. A solution of the variational principle, the minimum condition on the free energy functional at given ρ or μ, exists if and only if there exists a solution with $x \geq 0$ and $\rho_0 \geq 0$ for the unknown values in these equations. If we obtain the values of these two parameters then we uncover once again the explicit form of the quasi-free equilibrium states of the Bogoliubov model Eq. (4.53).

On the way to the solutions for these parameters, we verify that Eq. (4.56) is already satisfied and therefore that only the following two equations, in the variables x, ρ_0, remain to be considered and solved:

$$\rho_0 \left\{ x + \frac{1}{2} \int dk \left(\frac{\varepsilon_k + x}{E_k} \coth \frac{\beta E_k}{2} - 1 \right) \right\} = 0 \tag{4.64}$$

$$\frac{x + \mu}{v(0)} = \rho_0 + \frac{1}{2} \int dk \left\{ \frac{\varepsilon_k + x + \rho_0 v(k)}{E_k} \coth \frac{\beta E_k}{2} - 1 \right\} \tag{4.65}$$

Let us first consider the density constraint equation (4.65). Denote its right-hand side for shortness by $I(x, \rho_0)$. It is a strictly decreasing function of x. Furthermore, it can be solved for x. There exists a unique solution $x \geq 0$ if and only if ρ_0 belongs to the domain $D(\mu) = \{\rho_0 \in \mathbb{R}, I(0, \rho_0) \geq \frac{\mu}{v(o)}\}$. We will denote this solution by $x = f_\mu(\rho_0)$. As $\lim_{\rho_0 \to \infty} I(0, \rho_0) = \infty$, the function $f_\mu(\rho_0)$ is defined for all ρ_0 sufficiently large and also $\lim_{\rho_0 \to \infty} f_\mu(\rho_0) = \infty$. Moreover, if $\mu_1 < \mu_2$ and $\rho_0 \in D(\mu_1) \cap D(\mu_2)$, then $f_{\mu_1}(\rho_0) > f_{\mu_2}(\rho_0)$. For fixed $\rho_0 \geq 0$, there exists $\mu(\rho_0)$ such that $\rho_0 \in D(\mu)$ if μ is smaller than or equal to some μ_c; however, ρ_0 does not belong to $D(\mu)$ if $\mu > \mu_c$. In particular the so-called normal solution (zero condensate solution) $f_\mu(0)$ exists only for $\mu \leq I(0,0)v(0) = \rho_c(\beta)v(0)$ where $\rho_c(\beta)$ is the critical density of the free boson gas. If this occurs, $x = f_\mu(\rho_0)$ decreases near ρ_0 equal to zero, because here $\frac{\delta^2}{\delta\rho_0^2}I(x,\rho_0) < 0$.

Now we turn to the condensate equation (4.64). We remark that the expression in the curly brackets is decreasing in ρ_0 from a value strictly exceeding x (at $\rho_0 = 0$) to $x_{max} = \frac{1}{2}\int dk\, v(k)$ (at ρ_0 tending to infinity). Therefore, if $0 \leq x < x_{max}$ we can solve Eq. (4.64) for ρ_0 and obtain either $\rho_0 = 0$, or $\rho_0 = g(x)$ where g maps the interval $[0, x_{max})$ into the non-negative axis $[0, \infty)$ independently of μ, with $\lim_{x \to x_{max}} g(x) = \infty$ and $\min g(x) > 0$.

Further analysis yields that the variational principle has solutions $\rho_0 = 0$, or has both the parameters x and ρ_0 away from zero. Hence the phase transition is of first order because it is accompanied by a jump in ρ_0. For the condensate region ($\rho_0 > 0$) this means that also $x > 0$, and that there is a gap in the spectrum given by $\Delta E = [x(x + 2\rho_0 v(0))]^{1/2}$. Moreover we note that the spectrum E_k has a parabolic minimum at zero momentum and not a linear behavior required by the Landau criterion Eq. (4.1) for superfluidity. This reopens again the question about the search for a derivation and understanding of the phenomenon of superfluidity on the basis of this microscopic model. This is not the most exciting encounter if we want to use a Bogoliubov type of model explaining the phenomenon of superfluidity.

On the other hand, we know [7] that the linearity of the spectrum $E_k \approx \sqrt{\frac{\rho_0 v(0)}{m}}|k|$ near $|k| = 0$ is effectively obtained by applying the so-called Bogoliubov approximation. The latter consists in replacing the annihilation and creation operator $a_0^\#$ by a c-number, a complex number, namely by setting $a_0 = c\sqrt{V}$ in the two typical Bogoliubov terms of the model Eq. (4.53). Physically this approximation consists in neglecting, by a brute force ad hoc procedure, all the quantum fluctuations (see Chapter Eq. (6)) of the zero mode a_0. Let us therefore effectively apply now this approximation to our model Eq. (4.53) exactly as it was prescribed in the original work of Bogoliubov. It is important to apply the approximation only in the two Bogoliubov terms of the model, not in the others where the zero mode operators are also present. In any case this leads to the following new model, for which we realize that by this approximation operation we introduced a sort of self-consistency relation which is given by $\rho_0 = a_0^* a_0 = |c|^2$. The Hamiltonian of this model looks therefore as follows

$$H_V^{Bog'} = \sum_k \varepsilon_k a_k^* a_k + \frac{1}{2V}\sum_{k \neq 0} v(k)(a_k^* a_{-k}^* \rho_0 + \rho_0 a_k a_{-k})$$

$$+ \frac{1}{V}a_0^* a_0 \sum_{k \neq 0} v(k)a_k^* a_k + \frac{1}{2V}\sum_{k,k'} a_k^* a_{k'}^* a_{k'} a_k \qquad (4.66)$$

This model is again solvable and solved rigorously by using the same argumentations and performing the same technical steps as we used in the originally introduced model Eq. (4.53). The main difference between the model Eq. (4.53) and the model

Eq. (4.66) is that the condensate equation (4.55), applied to the first form of the model, is replaced by a much simpler equation,

$$\rho_0 x = 0; \text{ with } x = v(0)\rho - \mu \qquad (4.67)$$

It is clear that in the case of condensation, hence if $\rho_0 > 0$, this equation generates the solution $x = 0$. This implies directly the linearity of the spectrum in the zero momentum region. For this reason the model Eq. (4.66) meets the aspirations of understanding the phenomenon of superfluidity at a microscopic level. Also, we note on the other hand that now in the absence of condensation ($\rho_0 = 0$) the spectrum exhibits a spectral gap $\Delta E = x > 0$ as well. We should realize that here again we encounter an other model with an interaction able to create a spectral gap (see Eq. (4.2)).

These preceding efforts for a rigorous analysis of the Bogoliubov theory by means of the two models Eq. (4.53)and Eq. (4.66) teach us a great deal about the seriousness and the impact of neglecting the quantum fluctuations of the condensate particles. If we apply the approximation of Bogoliubov, we neglect these fluctuations and the spectrum becomes linear near zero momentum. If we take all the quantum fluctuations into account, the spectrum is parabolic in the neighborhood of zero momentum. Both situations are unsatisfying. As an overall conclusion we cannot but conclude that the challenge for obtaining a genuine microscopic Hamiltonian model remains. Such a theory would produce a full microscopic explanation of superfluidity on the basis of Bose-Einstein condensation. One particular way of posing the problem along the lines of the theory of Bogoliubov could be the following: It remains still an open question to identify which terms, present in the fully interacting two-particle boson Hamiltonian Eq. (2.4), or which mechanisms are responsible for the damping of these quantum fluctuations, suppressed by the Bogoliubov approximation. Of course another way of proceeding consists in discovering a brand new performing model which is completely independent from the Bogoliubov prescription—*a real challenge*.

4.7 Condensation in Two-body Fully Interacting Models

Rigorous proofs for the appearance of Bose-Einstein condensation in what we would call a realistic homogeneous systems has been an ongoing challenge for many decades. We should add that sometimes the word "realistic" does not only refer to the interaction term. For reasons of collegial correctness, we should mention that the word "realistic" includes also the presence of a one-particle spectrum of the type $\varepsilon_k = k^2/2m$, being continuous in the variable $k \in \mathbb{R}^d$, which is primarily obtainable after having used periodic boundary conditions. A renewed interest in this standing open problem is observed since the successful experiments with trapped gases of alkali metals, although these experiments are concerned with inhomogeneous boson systems. For these systems there are a number of rigorous results obtained within the framework of the Gross-Pitaevskii [71, 72, 134] equation, which we briefly touch in

Section Eq. (4.8). As a many-body boson system, the Gross-Pitaevskii approach is a mean-field approach.

At any rate, for fully interacting homogeneous systems, non-solvable models and with periodic boundary conditions Eq. (2.4) so far a general rigorous proof of Bose-Einstein condensation does not exist. In the physics literature the periodic boundary conditions come across as the physically most relevant ones. This appears to be a cultural phenomenon rather than something founded on deeper physical argumentation. However if one is open minded, why not, for other boundary conditions, and if one is interested in systems with genuine two-body interactions between the particles, we can find obtain results about the existence of condensation. For instance, many of us will be inspired by the observation that attractive boundary conditions [147, 31, 160], but also some weakly interacting systems (the theory of Bogoliubov for example), can create gaps in the one-particle spectrum. This gap can enhance the creation of boson condensation and therefore could enhance the occurrence of boson condensation for the interacting systems.

Indeed with this information in mind, we can adopt the following point of view: We start with a single-particle spectrum, $\varepsilon_k^{\Delta} = k^2/2m$ (for $k \neq 0$) and $\varepsilon_0^{\Delta} = -\Delta < 0$ (for $k - 0$), having a fixed non-trivial gap Δ in its spectrum. It is immediately verified that a free system with such a gap shows condensation for large total densities or equivalently for sufficiently low temperatures. We could then ask whether this gap makes the condensate stable enough for the addition of small but bona fide two-body interactions. We might anticipate that the condensate particles, which are energetically isolated from the excited states by the gap, can survive the switching-on of gentle interactions. We can also imagine that these fluctuations must be of a macroscopic size in order to cross the gap and lift the condensed particles out of the lowest energy state. This idea of considering a gap in the spectrum is not new. In his book [109] London attempted to introduce the gap on heuristic grounds to clarify some of the spectral properties of superfluid helium. This idea was already used for a microscopic model in [31] with van der Waals type of interactions. The idea is fully exploited in [99] where Bose-Einstein condensation is proven for systems with weak but genuine two-body interactions. The Hamiltonian of the system has the form:

$$H_V^{\Delta} = \sum_k \varepsilon_k^{\Delta} a_k^* a_k + \frac{1}{2V} \sum_{k,k',q} v(q) a_{k+q}^* a_{k'-q}^* a_{k'} a_k \tag{4.68}$$

where the two-body potential satisfies again the conditions $v(0) \geq v(q) \geq 0$ on the Fourier transform of the integrable super-stable potential.

In [99] it is proven in dimension $d \geq 3$ that for a fixed inverse temperature β and a chemical potential $\mu > v(0)\rho_c(\beta)$, where $\rho_c(\beta)$ is the critical density of the free boson gas, there exists a minimal strictly positive value Δ_{min} of the gap such that for all gaps $\Delta \geq \Delta_{min}$, the thermodynamic limit of the zero-mode occupation density is strictly positive. We can show explicitly that there exists zero mode condensation for the model Eq. (4.68) expressed by the inequality

$$\rho_0^{\Delta}(\beta,\mu) \equiv \lim_V \frac{1}{V} \frac{tr\, e^{-\beta(H_V^{\Delta} - \mu N_V)} a_0^* a_0}{tr\, e^{-\beta(H_V^{\Delta} - \mu N_V)}} > 0$$

This proves the existence of Bose-Einstein condensation for this type of bona fide interacting boson systems. This proof contradicts the possible idea that there is a no condensation possible when one adds gentle but genuine interactions to a system already showing condensation. Without reproducing all details of the full proof of this result, we should mention that the idea of the proof is based on the following items: We compare the condensate density of the full model with that of a special reference system, which is tuned by the given fully interacting model system. The reference system is not the free boson gas because that would immediately rule out the use of large values of the chemical potential. It is already remarked in [54] that even the mean field boson gas cannot be considered a perturbation of the free boson gas perturbed with the mean field interaction term. Instead we choose as a reference system a mean field boson gas that is indeed a perfect super-stable boson system. As a reference system, we use the mean field boson system obtained by taking the van der Waals limit of the given fully interacting boson system Eq. (4.68). This mean field limit features the Hamiltonian

$$H_V^{\Delta,mf} = \sum_k \varepsilon_k^\Delta a_k^* a_k + v(0) N_V^2 / V \tag{4.69}$$

Clearly the kinetic energy term is the same as in Eq. (4.68). The result about the relation between the condensate densities of the full model and the reference system is obtained by using essentially only convexity arguments of the thermodynamic functions. The first argument uses the convexity of the free energy with respect to the parameter Δ. The second argument uses the Peierls-Bogoliubov convexity inequalities which we reproduce here below (see also [169] Appendix D). It is by itself an interesting inequality which turns out to be useful in statistical mechanics at many other occasions.

Let H be any Hamiltonian, a self-adjoint operator on a Hilbert space such that $tr e^{-\beta H} < \infty$ for all $\beta > 0$.

Let g be any unit vector of the Hilbert space and $\{g_n\}_n$ an orthonormal basis of the space diagonalizing the Hamiltonian $H g_n = \sum_n E_n g_n$. Also, consider the decomposition $g = \sum c_n g_n$ with $\sum |c_n|^2 = 1$. The convexity of the exponential function immediately yields $(g, e^{-\beta H} g) = \sum |c_n|^2 e^{-\beta E_n} \geq e^{-\beta \sum |c_n|^2 E_n} = e^{-\beta (g, Hg)}$. Hence for any orthonormal basis we obtain the convexity inequality

$$tr e^{-\beta H} \geq \sum_n e^{-\beta (g_n, Hg_n)}$$

Consider now two self-adjoint operators H_1 and H_2 with the property: for $i = 1, 2$

$$|tr e^{-\beta H_i}(H_2 - H_1)| < \infty$$

Denote by ω_i, $i = 1, 2$, the Gibbs states determined by the Hamiltonians H_i, $i = 1, 2$ and defined by $\omega_i(A) = tr e^{-\beta H_i} A / tr e^{-\beta H_i}$. Consider the basis $\{f_n^{(2)}\}$ diagonalizing H_2 with eigenvalues $E_n^{(2)}$. Using the convexity inequality proved above, and once more the convexity of the exponential function, we obtain respectively

$$\frac{tr e^{-\beta H_1}}{tr e^{-\beta H_2}} = \frac{\sum (f_n^{(2)}, e^{-\beta(H_2+H_1-H_2)} f_n^{(2)})}{tr e^{-\beta H_2}}$$

$$\geq \frac{\sum e^{-\beta E_n^{(2)}} e^{-\beta(f_n^{(2)},(H_1-H_2) f_n^{(2)})}}{tr e^{-\beta H_2}}$$

$$\geq \exp\{-\beta \omega_2 (H_1 - H_2)\} \tag{4.70}$$

Hence we obtain the inequality

$$\frac{tr e^{-\beta H_1}}{tr e^{-\beta H_2}} \geq \exp\{-\beta \omega_2 (H_1 - H_2)\}$$

Interchanging the operators H_1 and H_2 we also obtain

$$\frac{tr e^{-\beta H_2}}{tr e^{-\beta H_1}} \geq \exp\{-\beta \omega_1 (H_2 - H_1)\}$$

We next use these two inequalities in combination to derive inequalities for the free energies of the systems $F_i = -\frac{1}{\beta} \ln tr \exp\{-\beta H_i\}$ with the Hamiltonians $H_i, i = 1, 2$. We immediately obtain the following bounds for the difference of the free energies of the two systems in terms of the expectation values of the energy differences:

$$\omega_2 (H_2 - H_1) \leq F_2 - F_1 \leq \omega_1 (H_2 - H_1) \tag{4.71}$$

This inequality is called the *Peierls-Bogoliubov inequality*.

The detailed proof of the existence of boson condensation mentioned above for the model Eq. (4.68) can be worked out using these bounds and arguments and is left as an exercise.

In [99] the proof is concentrated on the case of dimensions $d \geq 3$. However the result can also be shown to hold in dimensions $d = 1, 2$ by a slightly different and somewhat more elaborate argument. Nevertheless the essential ingredient for the existence of condensation remains the presence of the gap Δ in the one-particle spectrum, which enhances condensation. The main conclusion of this result should be that Bose condensation does exist for fully interacting systems. It shows that Bose condensation is not just a prerogative of a number of solvable toy models.

4.8 BEC in Traps

Since 1995 and the precise experiments with boson gases hold together in traps a considerable amount of research (see e.g. [135, 130, 69, 34]) has focused on these particular types of non-homogeneous boson systems. We do not give a complete record about the status concerning these kinds of systems, which are not invariant under the space translations. Despite the absence of space translation symmetry, the research in this domain evolves toward the understanding of these experiments on the basis of the standard theory of condensation for homogeneous boson systems. We could ask the question: What may be the sense of applying such a theory, which is basically built on homogeneity, to nonhomogeneous systems? Indeed the basic theme in this book is the discussion of homogeneous boson systems. Faced with this situation, it is reasonable that we limit ourselves to give here only a formal modest introduction to trapped bosons. In particular, we try to understand in which sense this non-homogeneous type of condensation can be understood as being part of the homogeneous conventional Bose-Einstein condensation theory, which is discussed in the rest of this book. What is the main problem in fitting these trapped systems within the conventional boson systems? Boson systems in traps, that is, in external confining potentials, make the system inhomogeneous in space. One of the basic ingredients of the definition of conventional BEC is directly linked to the homogeneity of the system. Therefore it is clear that the type of condensation for trapped boson systems can only be considered as a conventional condensation in some limiting situations wherein the system becomes "effectively" homogeneous-like. For systems behaving in this direction, we may be tempted to think about the homogeneous systems with the particular boundary conditions given by scaled external potentials (see Eq. (4.9)). We pay some attention to these systems in this chapter and try to see a possible closer connection with the trap situation.

Another point which catches our attention, is the fact that trapped boson condensation is detected for extreme dilute gases within such a trap. Dealing with homogeneous systems, the physical condition is large densities and therefore basically different. Is there a link between the limit of extreme dilution creating BEC in traps and the high densities needed to produce conventional condensation? How do we interpret these two completely opposite prerequisites for condensation? Both are difficult questions.

Moreover in spite of the extreme dilution in the trapped case, it seems also necessary to take into account interactions between the atoms to fit and/or to explain the experimental data for trapped gases. Also this is not simply understood. On the other hand the use of the Gross-Pitaevskii equation [71, 72, 134] seems to be an excellent tool to fit the experimental data.

Because of all these premises we should understand that reaching our goal will likely not be an easy matter. For all these reasons we must confine ourselves to a short pedestrian description of the problem we face.

4.8.1 Free Boson Gas in an Harmonic Potential

As always, everything starts with the free boson gas. The experimental realization of BEC achieved in atomic gases seems to be, in may cases, well approximated by the harmonic potential trap. For a review on the applications of this model see [135]. We place a free gas of boson particles in an external 3-dimensional harmonic potential trap given by the external field $V_e(x)$

$$V_e(x) = \frac{1}{2}(\omega_1^2 x_1^2 + \omega_2^2 x_2^2 + \omega_3^2 x_3^2).$$

where the ω_i are the harmonic frequencies. It is a quantum mechanics textbook solvable exercise to write the Hamiltonian for N bosons as

$$H_N = \sum_{i=1}^{N} \sum_{j=1}^{3} \omega_j \left(a_{j,i}^* a_{j,i} + \frac{1}{2} \right) ; \quad \text{for } \hbar = 1 ; \ m = 1$$

where the $a_{j,i}$ and their adjoint are boson creation and annihilation operators. The spectrum of the Hamiltonian is given by the energy values per particle

$$E_n = \sum_{j=1}^{3} \left(n_j + \frac{1}{2} \right) \omega_j ; \ n = (n_1, n_2, n_3) \in \mathbb{N}^3 .$$

It is clear that the lowest energy value corresponds to the case $n = 0$. The ground state wave function is given by the normalized function

$$\varphi(x) = \prod_{i=1}^{N} \varphi_0(x^i) , \ x^i = (x_{1,i}, x_{2,i}, x_{3,i}) \in \mathbb{R}^3$$

and

$$\varphi_0(x) = \left(\frac{\omega_0}{\pi} \right)^{3/2} \exp \left[-\frac{1}{2} (\omega_1 x_1^2 + \omega_2 x_2^2 + \omega_3 x_3^2) \right]$$

where $\omega_0 = (\omega_1 \omega_2 \omega_3)^{1/3}$ is the geometric average of the oscillator frequencies for the three directions. The ground state particle density distribution is given by the function

$$n(x) = |\varphi_0(x)|^2$$

It is a Gaussian with the size of the cloud fixed by the oscillator length scale $a_0 = \omega_0^{-1/2}$. The velocity or momentum distribution $\hat{\varphi}_0(k)$ is also a Gaussian centered around zero momentum with a size of approximately $a_0^{-1} = \omega_0^{1/2}$. All this is valid for this free gas and it is clear that the presence of interactions may change drastically the form of the Gaussian peak visible for the noninteracting case.

We continue the harmonic gas case and turn to the finite temperature situation. Consider the Gibbs grand canonical ensemble. The Gibbs state density of particles at inverse temperature β and chemical potential μ is given by

$$\left\langle \frac{N_N}{N} \right\rangle = \frac{1}{N} \sum_{n \in \mathbb{N}^3} \frac{1}{e^{\beta(E_n - \mu)} - 1}$$

where N_N is the total number operator for the N boson particles. Comparable with the reasoning used in the uniformly homogeneous boson gas Eq. (4.2), we distinguish the lowest energy E_0-term contribution from the rest and consider the particle density in the lowest energy level

$$\left\langle \frac{N_0}{N} \right\rangle = \frac{1}{N} \frac{1}{e^{\beta(E_0 - \mu)} - 1} .$$

For N tending to infinity, if it happens that the chemical potential μ tends to the value $E_0 = (\omega_1 + \omega_2 + \omega_3)/2$ with a first-order correction term (which is of the order of $1/N$), then the density of the lowest energy particles $\langle N_0 \rangle / N$ can become non-zero and finite. This lowest energy level gets a macroscopic occupation, which for this model can be interpreted as this system showing a phenomenon of Bose-Einstein condensation. We need to understand that there is some work to do in order to fill the gap between the prime microscopic system and this macro-phenomenon on the basis of the indicated specific behavior of the chemical potential as a function of the large number of particles.

In any case, if all this can be realized, then the critical density can be defined as follows, again in the spirit of what is done for homogeneous systems:

$$\rho_c(\beta,\mu)_N = \frac{\langle N_N \rangle - \langle N_0 \rangle}{N} = \frac{1}{N} \sum_{n \neq 0} \frac{1}{e^{\beta(\omega_1 n_1 + \omega_2 n_2 + \omega_3 n_3)} - 1} < \infty.$$

Clearly, this way of proceeding creates a formal scenario for the realization of boson condensation, which is very much comparable to that discussed in the homogeneous case (see the saturation argument in Section Eq. (4.2)).

It is instructive to note, that for trapped systems, there is no volume dependence and hence no singularities nor thermodynamics limits, which had a central role in the concept of condensation for homogeneous systems. Therefore condensation in trapped boson systems is and remains so far a finite particle problem in which, strictly speaking, no phase transition can be traced comparable to the homogeneous gas.

Apart from the ground state there are the excited states of the system which build up the critical density. For homogeneous systems it is obtained after having taken the thermodynamic limit. For these trapped systems, theoretical physicists are also looking for other limits doing the same job. Some types of semiclassical approximation limits are proposed for the treatment of the excited energy levels. This is a limit procedure by which the energy level spacing becomes smaller and smaller. For instance it can be realized by taking the limit N tending to infinity together with the oscillator parameter ω_0 tending to zero in such a way that the product combination $N\omega_0^3$ remains a finite constant. If the constant ρ stands again for the total density of particles we compute for the critical density

$$\rho_c(\beta,\mu) = \rho - \rho_0 = \int_0^\infty \frac{\rho(\varepsilon) d\varepsilon}{e^{\beta\varepsilon} - 1}$$

where $\rho(\varepsilon)$ is the density of states of the particles with energy ε, calculated from the spectrum E_n. The density of states in the thermodynamic limit for the free particle system is given by

$$\rho(\varepsilon) = \int dx dp \frac{1}{(2\pi)^3} \delta(\varepsilon - \varepsilon(x,p))$$

where $\varepsilon(x,p)$ is the one-particle (quasi-particle) energy, $\varepsilon(x,p) = \frac{p^2}{2m} + V_e(x)$ with $V_e(x)$ the external potential. For the free boson gas model ($V_e(x) = 0$) with periodic boundary conditions we obtain the well-known formula

$$\rho(\varepsilon) = \frac{V m^{3/2}}{\sqrt{2}\pi^2} \sqrt{\varepsilon}$$

For the harmonic oscillator trap model introduced above we obtain the function

$$\rho(\varepsilon) = \frac{\varepsilon^2}{2\omega_0^3}$$

Hence for the density of states we find a significant difference between the harmonically trapped free particle system and the homogeneous free boson gas. This explains the difference in temperature behavior of the critical densities for the two models. It explains the strong dependance of the thermodynamic behavior of the system on the shape of the external potential.

At this point it seems instructive to compare the result of the limit procedure described above with the rigorous results obtained for the free homogeneous boson gas in scaled external fields [136, 121]. The latter systems are approximately homogeneous systems in the sense that they become homogeneous only in the thermodynamic limit. Indeed, consider again the one-particle Hamiltonian Eq. (4.9) for one-dimensional intervals V of lengths L with periodic boundary conditions

$$h_L = -\frac{\Delta}{2m} + V_e\left(\frac{x}{L}\right) - \mu_L$$

where we take the external potential of the type: $V_e(x) = c|x|^\alpha$; $c, \alpha > 0$.

The corresponding dynamics α_t^V for finite volume V (see Eq. (5)) is given for any test function $f \in \mathscr{S}$ by

$$\alpha_t^V a^*(f) = a^*(e^{ith_L}f)$$

Because of this special form of the dynamics, we are referring to a solvable model (see Eq. (5.1)). For finite L the system is not space-translation invariant nor homogeneous precisely because of the presence of the external potential. But as L becomes larger and larger the potential becomes more and more effective only at the boundaries. The external potential becomes trivial within the bulk of the system. For this reason we sometimes speak about these models in terms of weak external-field boundary conditions. The system again becomes clearly homogeneous in the limit of L tending to infinity. The space translation symmetry is repaired. As we will see, the external potential maintains its effects only as a collective phenomenon.

In this model the position operator x and the momentum observable p evidently satisfy the usual canonical commutation relation $[x, p] = i1$.

We realize also immediately that in this problem the variable $y = x/L \in [-1, 1]$ becomes a relevant parameter of the system overtaking for a deal the variable x. Note also that we obtain at least formally the limit commutators

$$\lim_{L \to \infty} [y, p] = \lim_{L \to \infty} [y, x] = 0$$

The parameter y commutes with the position x and momentum observable p, and becomes in this limit an additional independent external parameter of the system. However it is always good to keep in mind the original physical meaning as a position variable. Furthermore, there is indeed a formal similarity between the limit L tending to infinity in the scaled external field model and the limit N tending to infinity while the combination $N\omega_0^3 = \rho$ is kept constant for the oscillator model. Comparing these two model limits, it is as $\omega_0 \simeq 1/N^{1/3}$ or alternatively as we introduce the variable $x/N^{1/3}$. In other terms it is as if the length parameter L is replaced by $N^{1/3}$.

In any case the limit L tending to infinity is a meaningful thermodynamic limit for the model Eq. (4.9), yielding the following exact results: The total density formula in the thermodynamic limit becomes

$$\rho = \rho_0 + \tilde{\rho}(\beta, \mu)$$

where

$$\tilde{\rho}(\beta, \mu) = \frac{1}{2\pi} \int dk \int_{-1}^{1} dy \frac{1}{e^{\beta\left(\frac{k^2}{2m} + V_e(y) - \mu\right)} - 1}.$$

The critical density is defined in the usual way to be $\tilde{\rho}_c(\beta) = \tilde{\rho}(\beta, 0)$

We can see that the critical density $\tilde{\rho}_c(\beta)$ is always finite if the power α of the potential $V_e(x) = c|x|^\alpha$ satisfies the condition $\alpha \leq 1$. Of course all these computations can be generalized to higher dimensions with the necessary changes to the conditions on the potentials.

If the potential is of the type that allows a finite critical density $\tilde{\rho}_c(\beta)$, then for all densities ρ larger than the critical one ($\rho > \tilde{\rho}_c(\beta)$) there is Bose-Einstein condensation expressed by $\rho_0 > 0$, where ρ_0 is again the density of the condensate. We can check that the limit chemical potential $\mu = \lim_L \mu_L < 0$ if $\rho < \tilde{\rho}_c(\beta)$ and $\mu = 0$ if $\rho \geq \tilde{\rho}_c(\beta)$.

Looking closer at the explicit expression of the density $\tilde{\rho}(\beta, \mu)$, which is quite different from that of the free boson gas with periodic boundary conditions, we find that the condensate should appear concentrated at the points (k, y) satisfying

$$\frac{k^2}{2m} + V_e(y) = 0.$$

Hence, if we choose the potential of the form $V_e(x) = c|x|^\alpha$, the condensate is in the ($k = 0$)-mode and at the spacial point situated in $y = 0$. Of course we can repeat these arguments for more general types of external potentials and compute that the condensation, if any, takes place in general at the spacial zeros of the external potential.

4.8.2 Interacting Bosons in Traps

In this section a formal description is given of what is called the Gross-Pitaevskii equation. For a much more complete and elaborate exposition including the applications, we refer to other works (e.g. [34]). We restrain the material to ground state ($T = 0$) considerations.

Consider a bona fide two-body interacting boson system with a potential v and add an external field potential V_e. The system is formally given by a Hamiltonian of the type:

$$H = \int dx\, a^*(x) \left(-\frac{\Delta}{2m} + V_e(x) \right) a(x)$$

$$+ \frac{1}{2} \int dx\, dx'\, a^*(x) a^*(x') v(x-x') a(x') a(x)$$

expressed in terms of the usual formal creation and annihilation operators satisfying the usual CCR-relations

$$\left[a(x), a^*(x') \right] = \delta(x - x') \; ; \; \left[a(x), a(x') \right] = 0.$$

The first term is the kinetic energy and the external field contribution. The second term is the two-body interaction energy. Let ω_∞ be a ground state of this system and introduce the canonical transformation, called the field translation Eq. (7.3), which map the a-variables to the b-variables,

$$a(x,t) = \varphi(x,t) + b(x,t)$$

where $a(x,t)$ is the time evolved annihilation operator under the dynamics determined by the Hamiltonian H. The function $\varphi(x,t)$ is taken to be the expectation value of the annihilation operator in the presupposed ground state ω_∞

$$\varphi(x,t) = \omega_\infty(a(x,t)) \in \mathbb{C}.$$

This function is called the order-parameter function and also sometimes the "(classical) wave function of the condensate." We should note that such a non-trivial order parameter $\varphi(x,t) \neq 0$ presupposes in fact that the gauge symmetry is broken for the ground state ω_∞. Of course it is a priori unclear (a) when such a ground state exists and (b) for which interactions v, and for which external potentials V_e? This problem of existence is even more serious having taken into consideration the fact that the system is not homogeneous. This is a problem far from being resolved.

Following the spirit and the interpretation of the one-particle correlation functions introduced in Chapter Eq. (2.3) for ergodic homogeneous systems, we identify also by generalization the function

$$\rho_0(x,t) = |\varphi(x,t)|^2$$

as the density of the condensate particles, which however, due to the presence of the external potential, now depends explicitly on the space position variables x. This property should be clear because of the non-homogeneity of the model system.

The new quantum variables $b(x,t)$ are in any case obtained from the a-variables by means of a canonical transformation such that they are again boson annihilation and creation operators satisfying the equal time canonical commutation relations

$$\left[b(x,t), b^*(x',t) \right] = \delta(x - x') \; , \; \left[b(x,t), b(x',t) \right] = 0.$$

The dynamical equation of the annihilation operator (Heisenberg equation of motion) is explicitly computed as

$$i\frac{\partial}{\partial t}a(x,t) = [H, a(x,t)]$$

$$= \left(-\frac{\Delta}{2m} + V_e(x) + \int dx'\, a^*(x',t)\, v(x-x')\, a(x',t)\right) a(x,t)$$

Now let us consider the effective pseudo-potential

$$v(x-x') = g\,\delta(x-x')$$

where the constant g is related to the scattering length a by the relation $g = \frac{4\pi}{m}a$. We then obtain

$$i\frac{\partial}{\partial t}a(x,t) = \left(-\frac{\Delta}{2m} + V_e(x) + g\,a^*(x,t)\,a(x,t)\right) a(x,t)$$

Now we make what is sometimes called the (semi-)classical approximation, in which we make the substitution of the operator $a(x,t)$ by the function $\varphi(x,t)$, the expectation value of the operator. In other words we forget about the presence of the quantum boson field $b(x,t)$ and its typical quantum fluctuations (see Eq. (6)). At any rate, we obtain a nonlinear Schrödinger-type equation for the so-called wave function of the condensate $\varphi(x,t)$

$$i\frac{\partial}{\partial t}\varphi(x,t) = \left(-\frac{\Delta}{2m} + V_e(x) + g|\varphi(x,t)|^2\right) \varphi(x,t)$$

This non-linear equation is called the *Gross-Pitaevskii(GP) equation* [71, 72, 134]. Its derivation is clearly a very formal one which does not claim any deeper understanding on the basis of any microscopic theory. It is based on a brute force mean field assumption and a semi-classical approximation. Recently an elaborate and rigorous derivation of the GP-equation has been given in [43] in the following sense: The authors start from a system of N weakly interacting bosons via a pair potential of the special form $N^{-1}a^{-3}v(x/a)$. They consider the corresponding BBGKW hierarchy in a limit consisting of the thermodynamic limit coupled with the limit of the scattering length parameter a tending to zero. As far as the thermodynamic limit is concerned, it coincides essentially with what is called the van der Waals or the mean field limit. The authors show that the hierarchy decouples, or has a factorized solution, if the GP-equation (the non-linear Schrödinger equation) is satisfied.

In other words this GP-equation describes a semi-classical boson liquid if we can show that this equation has a non-trivial solution. Exact results within this context concerning this problem can be found in [105, 106, 100].

It is a general belief that this equation has a wide range of validity if the scattering length a is small with respect to the average distance between the particles. Therefore this condition is expressed in terms of the particle density ρ by the inequality $\rho a^3 \ll 1$. In any case the GP-equation has been successfully used by a great number of authors to fit and analyze many of the experimental data about boson condensations in traps.

We can sometimes also entertain alternative derivations of the GP-equation, considering it as the Euler equation minimizing the energy density functional as a function of the condensate density function φ. In particular we consider the energy functional

$$E(\varphi) = \int dx \left(\frac{1}{2m}|\nabla\varphi|^2 + V_e(x)|\varphi|^2 + \frac{g}{2}|\varphi|^4 \right)$$

The GP-equation is obtained as the energy minimizing Euler equation given by

$$i\frac{\partial}{\partial t}\varphi(x) = \frac{\delta E(\varphi)}{\delta\varphi(x)} .$$

We can compare the status of the energy functional $E(\varphi)$ with the energy functional of the well-known Ginzburg-Landau theory, which has shown its utility as an important tool in the phenomenological studies of superfluidity and superconductivity.

5

Boson System Dynamics

The basic time evolutions or shortly the dynamics of Hamiltonian systems describe reversible dynamics in both classical Newtonian mechanics and quantum mechanics. Hence the basic time evolutions for boson systems reflect those of reversible dynamics.

Let us first consider the reversible boson system, paying special attention to its dynamics in the infinite volume limit. As in the preceding chapters, we must maintain a clear distinction between solvable and non-solvable models. Needless to say, understanding the dynamics from the microscopic point of view and studying the spectra of various model dynamics remain important endeavors in this field.

Many realistic systems manifest irreversible dynamics. Furthermore, the physical world comes across as irreversible and no existing universally valid equations of motion prescribe the universal basis of irreversibility. Irreversible behavior doggedly hides its true nature. Although the field has produced tangible results, no general basic theory has yet appeared.

The best known situation of such irreversibility is undoubtedly the case of irreversible dynamics driving an arbitrary state of the system to thermal equilibrium at a temperature determined by its surroundings (i.e., the heat bath of the system). Even here, a deeper understanding of this phenomenon and the knowledge of more precise properties remain important to research in statistical mechanics. The challenge is to unravel those specific dynamics possessing stationary states far from equilibrium. Herculean efforts to solve these problems date back to the 1960s in the context of quantum systems and, in the context of classical mechanics, even back to the works of Boltzmann and Gibbs (although following more or less the same patterns). Although the main problems remain unresolved, this research has produced a number of interesting and significant results. See [35, 2, 100, 154, 156, 112, 113, 39, 114, 44] for examples.

For these reasons we discuss in this chapter reversible dynamics and some aspects of irreversible dynamics; however, we will always maintain the distinction between them. Naturally, we will also continue to distinguish clearly between solvable and unsolvable dynamics.

A.F. Verbeure, *Many-Body Boson Systems*, Theoretical and Mathematical Physics, 109
DOI 10.1007/978-0-85729-109-7_5, © Springer-Verlag London Limited 2011

5.1 Reversible Dynamics

In ordinary quantum mechanics a reversible dynamic is determined by a self-adjoint Hamiltonian H, an operator that acts on a Hilbert space. In the Heisenberg picture one considers the unitary operators $\{U_t = \exp(itH) | t \in \mathbb{R}\}$. The time evolution of an arbitrary observable X at time $t = 0$ is obtained at time t by $X_t = U_t X U_t^*$. In the infinite volume picture, comparable with what we do when searching for equilibrium states, the dynamics of a boson system is determined by local Hamiltonians H_V, one for each finite volume V. These Hamiltonians represent the sum of the kinetic energy of the particles and their interaction energy, with both given as in Eq. (2.4) for boson systems. These Hamiltonians are defined as self-adjoint operators on the Fock space. We set the one-particle kinetic energies in the Hamiltonian equal to $\varepsilon_k = k^2/2m$ or equal to $k^2/2m - \mu$, depending on whether the canonical or the grand canonical ensemble describes the system. Isolated systems are described by time-independent Hamiltonians. We once again define the unitary operators $U_t^V = e^{itH_V}$ for all real numbers $t \in \mathbb{R}$ and for all finite volumes V. We should note that we set Planck's constant $\hbar = 1$ to simplify the formulae. The one-parameter family of unitary operators $(U_t^V)_t$ enjoys the following group properties:

$$U_t^V U_s^V = U_{t+s}^V, \ (U_t^V)^* = U_{-t}^V, \ U_0^V = 1;$$

for a dense set of vectors ψ of the Fock space $\mathfrak{F}$ we assume the property $\lim_{t \to 0} ||(U_t^V - 1)\psi|| = 0$. This expression is known as the *strong continuity of the one-parameter group of unitary operators* $(U - t)_t$. The time evolution is given again by the Heisenberg dynamical equation. Starting with an observable X at time $t = 0$, its time evolved equivalent is given by

$$\alpha_t^V(X) = U_t^V X U_t^{V*} \tag{5.1}$$

It is important to realize that the maps α_t^V for all t leave the canonical commutation relations of the boson algebra of observables invariant and are therefore all canonical transformations Eq. (7.3). For more general mathematical details on the dynamical maps we refer to the Appendix Eq. (7.1).

For infinitely extended many-body boson systems, the problem of the thermodynamic limit, $V \to \infty$, is again immediately posed. The problem is: How do we provide a mathematically acceptable meaning to the maps $\lim_{V \to \infty} \alpha_t^V$? We must prove the existence, in one mathematical sense or another, of a limit of operators. This is a topological problem; a topology must be chosen on the algebra of observables. At a first look, it comes across as convenient to take the norm topology, but that implies that the observable algebra must be equipped with a norm (see Appendix Eq. (7.1)). This holds true if we take for our boson systems the algebra of observables, which is generated by the Weyl operators Eq. (2.10). Much work has been done within this formulation [26]. But, even in this case we quickly meet technical difficulties (see e.g. [51]) which permit few applications to realistic physical models.

Instead of concentrating ourselves on these technical difficulties related to this limit, we describe a more restricted and more pragmatic procedure which leads also to exact results. Let us rewrite the finite volume dynamics explicitly as

$$\alpha_t^V(X) = \exp(it\delta_V)X = \sum_{n=0}^{\infty} \frac{(it)^n}{n!}\delta_V^n(X) \tag{5.2}$$

where δ_V is again the commutator operation introduced before, namely $\delta_V(X) = [H_V, X]$. We consider particular classes of states ω which satisfy the property that for a set of observables large enough and any chosen pair (A, B) of these observables, all the limit correlation functions of the following type do exist:

$$\omega(A\alpha_t(X)B) \equiv \omega(Ae^{it\delta}(X)B) \equiv \lim_V \sum_{n=0}^{\infty} \frac{(it)^n}{n!}\omega(A\delta_V^n(X)B) \tag{5.3}$$

This relation defines the limit derivation $\delta \equiv \lim_V \delta_V$, as well as the limit dynamics $\alpha_t = e^{it\delta}$ of the system. Of course this notation holds only if the volume limit can be given a meaning to exist at all. This should become clearer from the examples in the applications below. It is clear that these limits depend in general greatly on the choice of the set of state. The case that the limits exist only for a limited number of states is a real possibility. Below we describe the most practical and the most relevant choices for those states having the property of making these limits meaningful.

As explained before, each map δ_V is a linear map of the algebra into itself which satisfies all the properties of a *derivation*, namely δ_V is a linear map, and satisfies the Leibnitz rule $\delta_V(XY) = \delta_V(X)Y + X\delta_V(Y)$ and the property $\delta_V(X^*) = -\delta_V(X)^*$. These general properties should hold as well as for the limit map δ.

This presentation of the dynamics α_t in terms of a derivation indicates that these dynamics are in fact determined by the derivations δ, rather than by the Hamiltonians $(H_V)_V$. Indeed we should realize that not all derivations have a Hamiltonian by which they are defined or generated and that different Hamiltonians can define the same derivation.

The attentional among us should already have noted that, looking above at the EEB-equilibrium conditions Eq. (3.7), the equilibrium states also are essentially determined only by these derivations, which in principle can or cannot be generated by a set of local Hamiltonians (H_V). In fact the Hamiltonian plays in many instances only a secondary role, which is limited to considerations about details concerning physical interpretations of the energy function.

For boson systems the concrete problem of the limits Eq. (5.3) consists in arguing first that each δ_V makes sense for all observables A, B, X which are products of a finite number of creation and annihilation operators, all this considered under the state ω. Therefore all these derivations can also be transported or considered as actions on the set of all the correlation functions (see Eq. (2.7), Eq. (2.12)) of the state. In all cases the main point remains to argue that the property Eq. (5.3) remains valid in the limit V tending to infinity, and the same for the exponential α_t^V of δ_V for each value t of the time parameter. Now we proceed to the search for the existence of the limit dynamics for some specific boson model systems. First we mention a general but highly practical technical property concerning the fact that we deal with derivations in bosons systems. As δ_V is a derivation, it is sufficient to know and hence to compute the operation on the single creation and/or annihilation operators $a(f)^*$ or $a(f)$ for any test function $f \in \mathscr{S}$. Indeed δ as well as all the δ_V share the property

$$\delta(a^{\#}(f_1)...a^{\#}(f_n)) = \sum_{j=1}^{n} a^{\#}(f_1)...\delta(a^{\#}(f_j))...a^{\#}(f_n)$$

Finally we also note that not only the Heisenberg picture can be used in the definition of the dynamics; the generalized Schödinger picture (see Eq. (7.1)) also yields the possibility of formulating the problem of the dynamics. The Schrödinger picture is obtained by considering the dynamics as a mapping of the state space $\mathfrak{E}$ into itself.

Free Bose Gas

The local Hamiltonian of the free boson gas is again $H_V^{free} = \sum_k \varepsilon_k a_k^* a_k$, using periodic boundary conditions. Of course other boundary conditions can be considered equally well. Clearly the corresponding derivation is computed by $\delta_V^{free}(a_p^*) = [H_V^{free}, a_p^*] = \varepsilon_p a_p^*$. In particular this formula has the property that δ_V^{free} maps any creation operator or annihilation operator into itself multiplied by the corresponding one-particle energy. We immediately obtain the full dynamics

$$\alpha_t^{V,free}(a_p^*) = \sum_{n=0}^{\infty} \frac{(it\delta_V^{free})^n}{n!} a_p^* = e^{it\varepsilon_p} a_p^* \text{ and } \alpha_t^{V,free}(a_p) = e^{-it\varepsilon_p} a_p.$$

The limit V tending to infinity in the sense of Eq. (5.3), for each homogeneous state ω described by its (n,m)-point correlation functions, is straightforward because of the locality of the creation and annihilation operators. Hence the free boson gas dynamics is given by: for all $g \in \mathcal{S}$ and $a^*(g) = \int dk g(k) a_k^*$,

$$\alpha_t^{free}(a^*(g)) = a^*(e^{ith_{free}}g) \tag{5.4}$$

where h_{free} is the one-particle energy multiplication operator $(h_{free}g)(k) = \varepsilon_k g(k)$. This completes the definition of the limit reversible dynamics of the free boson gas. Naturally, the set of states for which the limit dynamics exists is extendable to a much larger set than the homogeneous states. For those included, this point is worthy of further exploration.

Mean Field Bose Gas

The local Hamiltonian with periodic boundary conditions for the mean field boson gas is again

$$H_V^{mf} = \sum_k \varepsilon_k a_k^* a_k + \frac{\lambda}{2V} N_V^2 \text{ where } N_V = \sum_k a_k^* a_k$$

The corresponding local derivation δ_V^{mf} becomes now

$$\delta_V^{mf}(a_p^*) = [H_V^{mf}, a_p^*] = \varepsilon_p a_p^* + \frac{\lambda}{2}(a_p^* \frac{N_V}{V} + \frac{N_V}{V} a_p^*)$$

Again we must compute Eq. (5.3) for some state ω or for some set of states and we have to consider the limit V tending to infinity. In the expression for $\delta_V^{mf}(a_p^*)$, we see the particle density operator $n_V = N_V/V$, therefore posing the question about the limit density $n = \lim_V n_V$. We note that n_V is an average operator which can be given a mathematical meaning for all homogeneous states (see [26]). However in this case, to get an explicit expression of the limit for these averages coming from the interaction, we may restrict already the choice of states. In the context of homogeneous boson systems it is natural to limit the choice of the states to the ergodic ones Eq. (2.26), leading immediately to the result

$$\omega(A\delta_{mf}(a_p^*)B) = \omega(A(\varepsilon_p + \lambda\rho)a_p^*B)$$

where $\rho = \lim_V \omega(N_V/V)$ is a real number representing the total density of particles in the state ω. Hence always under the state, as in the formula above, we get in this case

$$\delta_{mf}(a_p^*) = (\varepsilon_p + \lambda\rho)a_p^*$$

This result is essentially of the same type of action obtained above for the free boson gas, namely, it is a multiplication operator, now with the mean field function: $h_{mf}(p) = \varepsilon_p + \lambda\rho$. This expression is called the *one-particle mean field energy*. It is important to note here that this function depends heavily on the choice of the state ω via the density formula $\rho = \lim_V \omega(\frac{N_V}{V})$.

As for the free boson gas, we obtain once again the full dynamics of the mean field boson gas valid for all ergodic states acting on the creation operators $a^*(f)$ explicitly given by

$$\alpha_t^{mf}(a^*(f)) = a^*(e^{ith_{mf}}f) \tag{5.5}$$

The formal equivalence of the expressions Eq. (5.4) and Eq. (5.5) is at the origin of the popular sayings that the mean field is the same model as the free Bose gas. However we should recall our remark about the different thermodynamic behavior of the two models (e.g. the problem with the equivalence of ensembles for the free gas) being an argument in favor of the mean field model as a more appropriate model than the free one, as explained before (see [54] and [101]). We know already that this message has important consequences for the applications; in particular it can make perturbations made around the mean field model more reliable than perturbations around the free gas model. There is indeed an essential difference between the two models, namely their one-particle energies. For this reason, in the mean field case we do not speak of boson particles, but rather boson *quasi-particles*. The previously mentioned formal equivalence of the two models is sometimes used on the level of the local Hamiltonians. It means that one introduces for the mean field model the so-called local *effective Hamiltonian* of the mean field model. One introduces the Hamiltonian

$$H_V^{efmf} = \sum_k (\varepsilon_k + \lambda\rho)a_k^*a_k \tag{5.6}$$

Here we stress again that this effective Hamiltonian depends already on the chosen state ω, a property which we indicated with the letters index ef. We again compute the corresponding dynamics Eq. (5.3) defined by this effective Hamiltonian in the same previously chosen state. At this point the only meaningful state. As with the free gas, we can immediately confirm that the corresponding dynamics coincides with the previous one, namely α_t^{mf}, computed from the Hamiltonian H_V^{mf}. Here we should point out that indeed both Hamiltonians define the same dynamics when they act on local observables measurable in the same state. We can then ask whether such an equivalence holds for all interesting physical questions that can be asked about the two models H_V^{mf} and H_V^{efmf}?

As far as the search for equilibrium states is concerned, it follows indeed from Eq. (3.7) that H^{efmf} and H^{mf} yield the same equilibrium solutions. Nevertheless let us consider the difference of the two Hamiltonians and compute

$$
\begin{aligned}
H_V^{mf} - H_V^{efmf} &= \frac{\lambda}{2V}(N_V^2 - 2\rho V N_V) \\
&= \frac{\lambda}{2V}\{\int_V dx(a^*(x)a(x) - \omega(a^*(x)a(x)))\}^2 - \frac{\lambda}{2}\rho^2 V \quad (5.7)
\end{aligned}
$$

where again $\rho = \omega(N_V/V)$ for all finite V. The thermodynamic limit is taken keeping constant the density ρ. We find that up to a constant term, which is irrelevant for the definition of the dynamics, the difference between the original mean field Hamiltonian and its effective one is a variance-type operator of the local density operator $a^*(x)a(x)$ in the considered state ω. This means that working with the effective Hamiltonian is making the approximation of excluding in the mean field model all effects of the total density fluctuations. It is like making the approximation consisting of replacing the density operator by a c-number (see e.g. our discussion in Section Eq. (4.6)). This is why we should carefully note that working with the effective Hamiltonian yields a perfectly equivalent dynamics as far as it is acting on local observables; but the two models may show serious differences when studying (say) the dynamics of fluctuations or when looking at other observables which are non-local quantities. This point should become completely clear when reading Chapter Eq. (6), where the concepts of quantum fluctuation operators and their dynamics are introduced and studied as non-local but collective quantities.

Solvable Boson Models

The free boson gas and the mean field model are solvable models. Here we want to pay some special attention to the class of solvable models defined by the local Hamiltonians $\{H_V\}$ whose energy density for any ergodic state ω is expressed solely in terms of the one- and two-point functions of the state. In Chapter Eq. (4), a number of solvable models are considered mainly with the problem of looking for their equilibrium states. Here we look for the dynamics of this kind of models. In order to get a better idea about such models, we try to get an heuristic view about a general class of solvable models.

We limit ourselves to two-body interacting boson systems which are space homogeneous and gauge invariant. Hence we consider interaction energies as in Eq. (2.4), which are given by

$$H_V^i = \frac{1}{2V} \sum_{k,k',q} v(q) a_{k+q}^* a_{k'-q}^* a_{k'} a_k \qquad (5.8)$$

We look for models, hence Hamiltonians built up with contributions consisting of terms which appear in the sums of this formula and which lead to solvable models. Consider the interaction energy density $\lim_V \omega(H_V^i/V)$ for any ergodic state ω. Clearly taking into account that this energy density should not contain 3- and 4-point correlation functions, and therefore supposing that the 3- and 4-point truncated functions vanish, we find that the Hamiltonian of a solvable model can contain only the following type of terms:

$$\frac{v(0)}{2V} \left(\sum_k a_k^* a_k \right)^2 + \frac{1}{2V} a_0^* a_0^* \sum_{k \neq 0} v(k)(a_k a_{-k} + a_k^* a_{-k}^*) + \frac{1}{V} a_0^* a_0 \sum_{k \neq 0} v(k) a_k^* a_k$$

$$+ \frac{1}{2V} \sum_{k,k' \neq 0} v(k-k') a_k^* a_{-k}^* a_{k'} a_{-k'} + \frac{1}{2V} \sum_{k,k' \neq 0} v(k-k') a_k^* a_k a_{k'}^* a_{k'}$$

We realize that this general solvable interaction does contain as special cases all the solvable models treated in the previous chapter. For technical simplicity of the formulae we will not compute the dynamics of this general interaction Hamiltonian, but consider only a prototype model, namely the so-called *pairing model* [142, 143, 141] which contains the essential ingredients of generalizing the free and mean field models up to all the solvable type interaction Hamiltonian models. This boson pairing model is given as in [142, 143, 141] by the Hamiltonian

$$H_V^{pm} = \sum_k \varepsilon_k a_k^* a_k + \frac{v}{2V} N_V^2 - \frac{u}{2V} Q_V^* Q_V \qquad (5.9)$$

where $Q_V = \sum_k \lambda(k) a_k a_{-k}$ and λ is a real L^2-function satisfying $\lambda(-k) = \lambda(k)$ and $\lambda(0) = 1 \geq \lambda(k) \geq 0$; u and v are positive constants with $v - u > 0$, implying that the Hamiltonian describes a super-stable system. We leave the proof of this property as an exercise. It is again important to note that N_V/V and Q_V/V, in the limit V tending to infinity, are space averages. Indeed we also have $Q_V/V = (1/V) \int_V dx\, a(x) \int_V dx\, \widehat{\lambda}(x-y) a(y)$.

The model is mainly conceived to study the joint appearance of boson condensation ($\rho_0 > 0$) and of non-trivial pairing order that is $s(k) = \omega(a_k a_{-k}) \neq 0$. The latter one is called the pairing order parameter.

Again we define the dynamics for some state ω, which we take again ergodic, and follow the prescription of the definition Eq. (5.3). The ergodicity of the state guarantees again the existence of space average operators, which turn out to be multiples of the unit operator. We assume of course that our state has a finite total density of particles ρ and a finite value $q = \lim_V \frac{1}{V} \omega(Q_V)$ for the total pairing density. A direct application of the definitions and the proof of Eq. (5.3) begins again with the computation of the commutators

$$\delta_V^{pm}(a_p^*) = [H_V^{pm}, a_p^*] = (\varepsilon_p + \rho v) a_p^* - u\lambda(p) \overline{q} a_{-p}$$
$$\delta_V^{pm}(a_p) = [H_V^{pm}, a_p] = -(\varepsilon_p + \rho v) a_p + u\lambda(p) q a_{-p}^* \qquad (5.10)$$

For simplicity we consider q real or $\bar{q} = q$; the phase of q can be restored once finished by a suitable gauge transformation. Contrary to the free and mean field cases we do not have a simple multiplication function for the action of the commutator of the Hamiltonian on a creation/annihilation operator. We get a linear combination of the latter ones. However the important feature is precisely that the commutators are still linear in the basic boson operators. Therefore we can remedy this fault by performing a Bogoliubov canonical transformation Eq. (7.3) τ of the type used in Eq. (2.6), namely

$$a_p^* = \tau(\tilde{a}_p^*) = \tilde{a}_p^* \cosh \alpha_p + \tilde{a}_{-p} \sinh \alpha_p$$
$$\tilde{a}_p^* = \tau^{-1}(a_p^*) = a_p^* \cosh \alpha_p - a_p \sinh \alpha_p \qquad (5.11)$$

where α is a symmetric function of the momentum variable p determined by the formula:

$$\tanh 2\, \alpha_p = \frac{u\,\lambda(p)\,q}{\varepsilon_p + \rho\, v}$$

We recall that the creation and annihilation operators $(\tilde{a}_p, \tilde{a}_p^*)$ generate all observables as they are generated by the (a_p, a_p^*). A simple straightforward computation yields

$$\delta^{pm}(\tilde{a}_p^*) = \lim_V [H_V^{pm}, \tilde{a}_p^*] = (\varepsilon_p + \rho v - uq\lambda(p) \tanh \alpha_p)\tilde{a}_p^* \equiv E_p \tilde{a}_p^* \qquad (5.12)$$

which defines also the one-particle energy function E_p. In terms of the new generators $(\tilde{a}_p, \tilde{a}_p^*)$ of the observables, the commutator with the Hamiltonian is again a multiplication operator now with the function E_p of p. The function depends again heavily on the state which is chosen in the volume limit procedure. With the exception of the free boson gas, this situation is generic for nearly all solvable models. Clearly the total dynamics is again given by

$$\alpha_t^{pm}(\tilde{a}(f)) = \tilde{a}(e^{ith^{pm}} f) \text{ where } (h^{pm} f)(k) = E_p f(p)$$

The original creation and annihilation operators appearing in the original Hamiltonian Eq. (5.9) are sometimes called the operators of the *bare particles*, whereas the new operators are called the creation and annihilation operators of the *quasi-particle*. Equation (5.12) demonstrates that formally the system behaves like a free boson model but that very property is only recovered in terms of the new quasi-particles. We can again compute the *effective Hamiltonian* for this model as we did for the mean field model. For the pairing model we obtain

$$H_V^{ef\,pm} = \sum_k E_k \tilde{a}_k^* \tilde{a}_k$$

It turns out that the important feature of the effective Hamiltonian, as of all solvable Hamiltonian models, is that it is of the type of a free boson gas, bilinear in the creation or annihilation operators of the bare particles or of the quasi-particles. This is the basic characterizing property of all solvable boson model Hamiltonians. Note at the same time that the original Hamiltonian of a solvable model need not be necessarily bilinear in the bare creation and annihilation operators. The pairing model, and the Bogoliubov model are examples of such models. For these models the difference between the original model Hamiltonian and its effective Hamiltonian is again of the type of a quantum fluctuation, a property which we detected already for the mean field model.

Non-solvable Boson Models

The essential property of a non-solvable boson model is the following. Its Hamiltonian or its possible effective Hamiltonian on the basis of Eq. (5.3) for an ergodic state, is not expressed as a bilinear expression in some set of creation and annihilation operators. We can formulate this property in many different forms. Typical for a non-solvable model is that the commutators of the type $\lim_V [H_V, a(f)]$ (i.e. of the Hamiltonian with any creation or annihilation operator) contain terms which are products of two or more creation and/or annihilation operators. Taking the expectation values for any ergodic state of these commutators (see Eq. (5.3)), we see that the expression does not reduce to combinations of expectations in terms of one- and two-point functions. Stated otherwise, such commutators generate non-trivial higher-order product terms in these operators. If such is the case, applying a second commutator with the Hamiltonian $[H_V, [H_V, a(f)]]$ generates again new terms with again higher-order products in the creation/annihilation operators, and so on.

To define the full dynamics we must compute all these commutators with the Hamiltonian up to infinity. The dynamical equations written out in terms of the correlation functions yield an infinite set of coupled equations of infinitely many different correlation functions. The fact that the general Hamiltonian boson model H_V in Eq. (2.4) for a non-trivial two-body potential v is in general non-solvable, is straightforwardly checked by computing these commutators and by checking the presence of this infinite set of correlation functions in the dynamical equation.

For these general non-solvable boson systems, how do we even provide a reasonable mathematical meaning to the dynamics of such systems? Do the dynamics exists at all for such a general boson model and if it exists in which mathematical sense? In the mathematical physics literature we can find a number of results about this topic. We would find some arguments proving the existence of the dynamics for some general two-body interacting systems. However, few results of this kind appear. (See [26].) The known arguments are mostly worked out within the scope of a GNS-representation space of one or more states where the problem is transposed into a Hilbert space problem. All these type of proofs are in any case always, mathematically speaking, highly technical. Another technique dealing with this problem uses the theory of perturbations (see e.g.[26, 2]). In view of the scope of this book all these arguments are technically too much involved to be discussed in all details here. On the other hand we should note that there are few exact results known about the non-existence of the dynamics of such general models. At least this situation may be a good reason to skip further discussions about this problem and to continue with what is better known and what is technically more transparent.

5.2 Irreversible Dynamics

As already mentioned, the basic microscopic models for the dynamics of physical systems are given by reversible models. Essentially, we have the Newton model for classical systems and the Schrödinger-Heisenberg model for quantum systems. As mathematical constructions they share an exceptional degree of coherence and as physical theories they are exceedingly applicable. Together with their relativistic complements they come over as universal physical theories or models explaining all reversible phenomena.

In order to describe and explain irreversible phenomena the situation is much less clear. So far there is no universally applicable theory, not for classical nor for quantum irreversible phenomena. On the one hand there are many ad hoc dynamical phenomenological theories describing fairly well directly macroscopic phenomena. On the other hand considerable work has been performed to derive such irreversible or dissipative dynamics from the basic reversible equations of motion in special limit situations. In this sense many models have been constructed with a limited or at least a variable range of applicability. This situation is a consequence of the lack of universal understanding, and hence of formulation, of the genuine physical principles behind irreversibility. The question can be raised whether such principles have any chance of being realistic or to exist at all. In the last few decades a great deal of activity has arisen with the hope of making real progress in this interesting area of physics.

Faced with this situation, we limit ourselves here to a very special class of irreversible evolutions or dynamics. We restrict the discussion to those physical processes describing rather successfully the phenomenon of evolution towards equilibrium in physics. In particular this class of dynamics has been understood fairly well on the basis of microscopic reversible models. These are the result of considering the irreversibility phenomenon as the result of special approximations which are usually referred as the *weak coupling limit theory* [35].

Referring to Appendix Eq. (7.2) (see also [2] Chapter 8), any such irreversible dynamics will be given by a *quantum dynamical semigroup* $(\alpha_t)_{t\in\mathbb{R}^+}$ of linear maps of the algebra of observables into itself such that $\alpha_t\alpha_{t'} = \alpha_{t+t'}$ for all non-negative real values t,t' of the time parameter, with the additional conditions that $\alpha_t(1) = 1$ and that α_0 is the identity map.

In general, in order to get decent dynamics $(\alpha_t)_t$ of a physical system it is necessary and sufficient that, for all values of the time parameter $t \geq 0$, each state ω is mapped by the dynamics into a state $\omega \circ \alpha_t$ again. We should convince ourselves that this is guaranteed by the conditions that each particular map α_t is a linear, unity preserving, and positive map of the algebra of observables into itself. In particular the last property means that $\alpha_t(X^*X) \geq 0$ for any observable X.

There are many semigroups of this type. But the whole set of these semigroups is far from being completely classified. Therefore we restrict the discussion to the special class of these semigroups, namely the *Markovian dynamics*. They are characterized as follows: We assume that the map $t \to \alpha_t$ is regular enough such that α_t has the following structure: $\alpha_t = \exp(tL)$ for all $t \geq 0$. Hence these dynamics (α_t) are describable in terms of the map L, called the *generator* of the dynamics or of the time evolution. The study of such type of evolutions is done by means of the study of their generators L. Note that for the reversible systems the corresponding generators of the dynamics are given by the derivation or commutator maps of the type $L = \delta = i[H,.]$ where $H = H^*$ is a self-adjoint operator called the Hamiltonian of the system. Now looking to the non-reversible ones, we can ask the following questions: Which are the possible explicit forms of L? Which are the specific properties of such a generator L sufficient to meet the conditions for generating such a dynamical semigroup?

We limit ourselves to the type of dynamics which are also introduced in the Appendix Eq. (7.2). The most general generator of a dynamics, which can contain a part responsible for a reversible as well as for a dissipative dynamics, is given by the linear map L of the algebra into itself. In the physics literature this generator is usually called the *Liouvillian or Liouville operator*. It consists of two parts and of the form: $L = L_H + L_D$. Here, L_H is the Hamiltonian or the conservative (i.e., reversible) part of the generator L. It is a derivation mostly defined by a given Hamiltonian H in the form $L_H = i[H,.]$. On the other hand, L_D is the dissipative part of the generator given by the Lindblad [107] expression, which states that there exists any observable X such that L_D is the action

$$L_D = [X^*,.]X + X^*[.,X] \tag{5.13}$$

and all the positive linear combinations of such mappings. This means, for all $\lambda_i \geq 0$ and observables X_i, $L - D$ is of the form:

$$L_D = \sum_i \lambda_i \{[X_i^*,.]X_i + X_i^*[.,X_i]\} \tag{5.14}$$

The mathematical properties of this type of maps are discussed in the Appendix Eq. (7.2). We discussed already the reversible dynamics in the preceding section. Therefdore we can restrict our discussion here to generators of the type $L = L_D$, i.e. retaining only the dissipative or strict dissipative part of the generator.

Before discussing the applications to boson systems, we prove a general physical property of this special type of dynamics. In order to make everything more transparent we consider again the case of finite matrix systems. In this case the algebra of observables is given by $\mathscr{A} = M_n$, the $n \times n$ complex matrices. First we consider a Hamiltonian $H = H^* \in M_n$, with $H\varphi_k = \varepsilon_k\varphi_k$, the ε_k are the eigenvalues, and the φ_k are the eigenvectors of H forming an orthonormal basis of $\mathbb{C}^n$. Denote in the Dirac notation, $E_{k,l} = |\varphi_k><\varphi_l|$, the partial isometries mapping the vector φ_l into the vector φ_k. Consider all the following generators of the type as described above: For all k,l

$$L_{k,l} = e^{\beta(e_k - e_l)/2}\{E_{k,l}[.,E_{l,k}] + [E_{k,l},.]E_{l,k}\} + e^{\beta(e_l - e_k)/2}\{E_{l,k}[.,E_{k,l}] + [E_{l,k},.]E_{k,l}\}$$

Each of these generators determines a dynamics $\alpha_t = exp\{tL_{k,l}\}$. Let us now consider the following statement: *Suppose that $\omega(.) = tr\rho.$ is a state that is time invariant for all these dynamics, which means that $\omega \circ L_{k,l} = \frac{d}{dt}\omega \circ \alpha_t |_{t=0} = 0$ holds for all k,l. We show that the state ω is necessarily the canonical Gibbs state.* In order to prove this property, consider first the case $k \neq s$. From time invariance we obtain $\omega(L_{k,k}(E_{k,s})) = 0$. A direct computation yields $L_{k,k}(E_{k,s}) = -2E_{k,s}$ and we get $\omega(E_{k,s}) = (\varphi_s, \rho\varphi_k) = 0$, which means that the density matrix ρ of the state ω is diagonal in the basis diagonalizing the Hamiltonian H.

For $k \neq l$ we also compute:

$$L_{k,l}(E_{k,k}) = -2e^{\frac{\beta}{2}(E_k-E_l)}E_{k,k} + 2e^{\frac{\beta}{2}(E_l-E_k)}E_{l,l}$$

Applying again the state to this relation and using again the time invariance of the state yields

$$\omega(E_{k,k})e^{\beta E_k} = \omega(E_{l,l})e^{\beta E_l}$$

which expresses in particular that both sides of the equality are independent of the indices k or l. Hence $\omega(E_{k,k})e^{\beta E_k} = \lambda$, a real number independent of the indices. On the other hand, normalization of the state yields $\omega(1) = \sum_k \omega(E_{k,k}) = tr\rho = 1$, implying that $\lambda = 1/tr e^{-\beta H}$. Hence altogether we get

$$(\varphi_k, \rho\varphi_l) = \delta_{k,l}\omega(E_{k,l}) = \delta_{k,l}\frac{e^{-\beta E_k}}{tr e^{-\beta H}}$$

yielding that the state ω is the canonical Gibbs or equilibrium state $\omega(.) = \omega_\beta(.) = tr e^{-\beta H}(.)/tr e^{-\beta H}$ for the system H. This proves the announced statement.

Note the following: Consider any time-limit state ω, that is, a state of the type $\omega = \lim_{t\to\infty} \omega' \circ \alpha_t^{k,l}$, with $\alpha_t^{k,l} = e^{tL_{k,l}}$ for all k,l where ω' is some arbitrary state. This time-limit state ω, if it exists, is invariant for all these dynamics. As a consequence of the above statement, it is an equilibrium state. This property is at the origin of what is meant by these dynamics being suitable to describe the *evolution or approach to equilibrium*.

Before proceeding to the applications in boson systems of these type of irreversible dynamical maps we mention another important property of the following generator: $L : A \in M_n \to L(A)$

$$L(A) = \sum_{i,j} e^{-\frac{\beta}{2}(\varepsilon_i-\varepsilon_j)}\{E_{ji}[A, E_{ij}] + [E_{ij},]E_{ij}\} \tag{5.15}$$

We can check straightforwardly that for each pair A,B of observables and for the Gibbs state ω_β using explicitly its property $\omega_\beta(E_{kl}) = \delta_{k,l}e^{-\beta \varepsilon_k}/tr e^{-\beta H}$, holds

$$\omega_\beta(L(A)B) = \omega_\beta(AL(B)) \tag{5.16}$$

Looking at Eq. (5.15), we can spot immediately the symmetry mapping E_{ij} into E_{ji} and the inverse; E_{ij} describes the transitions from φ_i, energy ε_i eigenvector, to φ_j, eigenvector corresponding to ε_j. In other words the transition $\varepsilon_i \to \varepsilon_j$ is realized by the matrix unit operator E_{ij}.

The transition probability D_{ij} in Eq. (5.15) is identified to be $D_{ij} = \lambda e^{\frac{\beta}{2}(\varepsilon_i-\varepsilon_j)}$ where λ is a positive normalization constant and satisfies the symmetry property

$$D_{ij}\, \omega_\beta (E_{jj}) = D_{ji}\, \omega_\beta (E_{ii}) \tag{5.17}$$

The properties Eq. (5.16) and Eq. (5.17) are equivalent and are called the *detailed balance property of the generator L* or of its corresponding dynamics $\{e^{tL}|t \geq 0\}$, with respect to the Gibbs equilibrium state ω_β.

The form Eq. (5.16) expresses more directly the following mathematical symmetry property: If we consider again the scalar product $(A,B) = \omega_\beta (A^*B)$ on the observables algebra, then Eq. (5.16) reads as $(L(A),B) = (A,L(A))$ for all A and B. It expresses the symmetry (self-adjointness) of the generator L with respect to the scalar product defined by the Gibbs state. The form Eq. (5.17) of detailed balance is the more popular one in the physics literature.

Now we proceed to the applications of Lindblad generators in boson systems, and make again a explicit distinction between the solvable and non-solvable models.

Solvable Boson Models

In the reversible case the solvable models are obtained by means of Hamiltonians, which are effectively maximal bilinear in the creation and annihilation operators. For the irreversible dissipative case solvability is defined analogously in terms of the operators X, defining the generator L_D Eq. (5.13). The system dynamics is solvable for all choices of X being linear combinations of creation and annihilation operators. This corresponds to generators L_D being bilinear in these operators. Analogous as in the reversible case, we can take first the operators X equal to $X = a_p$ or $X = a_p^*$ and consider the corresponding generator. More general positive linear combinations of these generators can be considered. Explicitly we consider for all $\lambda (k) \geq 0$ the generator $L \equiv L_D$ given by

$$L = \sum_k \lambda (k)\{[a_k^*,.]a_k + a_k^*[.,a_k]\} \tag{5.18}$$

One computes again the thermodynamic limit following the prescription of Eq. (5.3) for all $t \geq 0$. From this point on, the situation closely reflects the reversible case. For the action of the generator on the annihilation observable a_p as well as on the creation observable a_p^*,

$$L(a_p) = -\lambda (p)a_p \text{ and } L(a_p^*) = -\lambda (p)a_p^*$$

The result of the action L being the same for the annihilation operators as for the creation operators, the generator of the dynamics is again reduced to a multiplication operator, this time by the non-positive function $-\lambda (p)$. The exponentiation of the operator L, yielding the dissipative dynamics $\alpha_t^D, t \geq 0$ is again straightforwardly obtained to be

$$\alpha_t^D (a_p) = e^{-t\lambda (p)}a_p$$

For a more sophisticated mathematical treatment of this model we refer to [161]. For a number of applications in physics of these types of dynamical semigroups we can consult the reference [3].

At this point it is important to remind ourselves that any reversible dynamics of the type $\{\alpha_t(.) = \exp(it[H, .])\}$ is a one-parameter set of canonical transformations Eq. (7.3). Indeed the canonical commutation relations are left invariant under all reversible time evolutions. Expressed in formulae this means for all p, p' and t holds

$$[\alpha_t(a_p), \alpha_t(a_{p'}^*)] = \delta_{p,p'} \text{ and } [\alpha_t(a_p), \alpha_t(a_{p'})] = 0$$

On the other hand this property does not hold anymore for dissipative dynamics. Using the above computation we find for any non-trivial multiplication function $\lambda(p) > 0$:

$$[\alpha_t^D(a_p), \alpha_t^D(a_{p'}^*)] = e^{-t(\lambda(p)+\lambda(p'))}\delta_{p,p'} \neq \delta_{p,p'} = \alpha_t^D([a_p, a_{p'}^*])$$

The lack of the conservation of the canonical commutation relations under dissipative evolutions leads in general to technically more involved algebraic computational work than in the reversible case. The literature contains many applications of this type of dynamics. For the field of photon physics we refer again to [3] and references therein. In [145] this type of dynamics is used to compute rigorously the temperature critical exponent of the susceptibility in the neighborhood of the critical point of the free boson gas condensation.

Non-solvable Boson Models

As in the reversible case, the non-solvable models are produced by taking the observable X defining the generator L Eq. (5.13) in such a way that it contains terms non-linear in the creation and annihilation operators. In that case again the commutators appearing in the expression for the generator L applied to a creation or annihilation operator generate terms which contain higher order products in the creation/annihilation operators than the linear terms, exactly as it is in the reversible case. Consecutive application of this generator becomes again a problem of infinite order, more or less analogous to the reversible case. On the other hand, it may happen in the non-reversible case that the strict dissipativity property of the generator is helpful in proving matters. In particular this strict dissipativity property may be helpful in proving the exponentiation of the generators. In other words it may help in proving the existence of the dynamics. Compared to the hamiltonian case, the strict dissipativity of the Liouville generator L_D offers here a supplementary asset.

6

Quantum Fluctuations and Bosonization

6.1 Preliminaries

Statistical physics is dealing with systems characterized by the fact that they have many degrees of freedom. One of the main problems consists of finding procedures for the extraction of the relevant physical quantities out of these extremely complex systems. We are faced with the problem of finding relevant reduction procedures which map the complex systems onto a simpler, tractable model at the price of introducing elements of uncertainty. Therefore probability theory is the natural mathematical tool in statistical physics. Since the early days of statistical physics, in classical (Newtonian) physical systems it is natural to model the observables by a collection of random variables acting on a probability space. Kolmogorovian probability techniques and results are the main tools in the development of classical statistical physics. A random variable is usually considered to be a measurable function with an expectation (state) given by an integral with respect to a suitable probability measure. Alternatively, a random variable can also be viewed as a multiplication operator by the associated function. Different random variables commute as multiplication operators. For this reason we speak of a commutative probabilistic model.

Looking at genuine quantum systems, in a number of cases, the mentioned procedure leads to commutative probabilistic models, but there are also the realms of physics where quantum non-commutative probabilistic concepts are unavoidable. Macroscopic quantum effects are really observed. Typical examples of such areas of physics are quantum optics, low temperature physics and ground state physics such as relativistic and non-relativistic quantum field theory. During the last half a century physicists have been developing more or less heuristic methods to deal with the manifestations of fluctuations with a typical quantum nature. The last thirty years have seen efforts to formulate also the mathematical foundations of such theories. Within this context even a notion of quantum probability was launched as a branch of mathematical physics and of pure mathematics [33, 144, 60].

This chapter aims to review briefly a few selected rigorous results concerning non-commutative or quantum-limit theorems with direct applications to quantum statistical mechanics. This item is chosen because of its close relation to concrete

A.F. Verbeure, *Many-Body Boson Systems*, Theoretical and Mathematical Physics,   123
DOI 10.1007/978-0-85729-109-7_6, © Springer-Verlag London Limited 2011

problems in statistical physics where we aim to understand the macroscopic phenomena on the basis of the microscopic structure. Therefore a precise definition or formulation of the notions of microscopic and macroscopic systems should be of prime importance. A common physicist's belief is that the macroscopic behavior of a system in the thermodynamic limit is described by a reduced set of macroscopic quantities [70, 152]. Examples include, amongst others, the average densities of particles, the average energy, momentum, magnetic moment, and so on. Analogous to the microscopic quantities, the macroscopic observables can be considered as elements of an algebra of observables; macroscopic states of the system could be states on the latter one. The main problem is to conceive and construct the precise structures and the mathematical procedures to go from a given microscopic system to its macroscopic systems. We will be permanently confronted with this process, which is sometimes expressed simply by the slogan phrase "from micro to macro".

A well known first example of a classical macroscopic system is the one based on the algebra of observables, the so-called set of *observables at infinity* [97], containing the spacial averages of all local micro-observables of a classical or quantum microsystem. Being explicit, for any local observable A we consider the macro-observable

$$A_\omega = \lim_{V \to \infty} \frac{1}{V} \int_V dx \, \tau_x A$$

where the limit $V \to \infty$ is the (weak operator) limit in the homogeneous micro-state ω, that is, for each pair of observables B, C: $\omega(BA_\omega C) = \lim_V \omega(B\{\frac{1}{V} \int_V dx \, \tau_x A\}C)$. The obtained limit operators A_ω are the result of the law of large numbers in probability. The algebra generated by these limit-observables $\mathscr{A}_\omega = \{A_\omega | A \in \mathscr{A}\}$ is an abelian algebra of observables of a macroscopic system. This algebra can be identified with an algebra with point-wise product of measurable functions for some measure which is also called a macroscopic state. Hence at the macro-level of the averages, the commutative probability always works regardless of whether the original micro-system is classical or quantum. So far for the law of large numbers which we interpreted as a procedure for going from "micro to macro" valid also for quantum systems the case in which we are most interested in this book.

This chapter also attempts to describe another analogous mapping from "micro to macro" for a different type of volume scaling, namely the scaling of fluctuations. More explicitly, for any local observable $A \in \mathscr{A}$, we consider the following limit operator, where compared with the averages the volume V is now replaced by the square root of the volume,

$$F(A) \equiv \lim_V \frac{1}{\sqrt{V}} \int_V dx \, (\tau_x A - \omega(\tau_x A)) \tag{6.1}$$

The problem consists in characterizing again the limit quantity $F(A)$, called the *fluctuation operator*, as an operator acting on a Hilbert space and to specify the type of limit as well as the mathematical character of the set of fluctuations. In classical probability theory this limit is handled by the central limit theorem. The central result which we want to discuss is the fact that this macro-system of quantum fluctuations has all the properties of a boson system for all micro-systems having the usual locality property. This result can be seen as remarkable. It can be considered as a quantum probabilistic basis for the emergence of the boson canonical commutation relations. Already this interpretation can be considered as a firm basis for treating the theory of quantum fluctuations in a book on boson systems.

Performing this quantum central limit theorem, we note immediately that not all locally different microscopic observables necessarily yield different fluctuation operators. Therefore the quantum central limit theorem realizes also a well defined procedure for the physical notion of coarse graining. It realizes a reduction procedure which is handled by the mathematical notion of an equivalence relation on the microscopic observables yielding the same fluctuation operator.

The material of this chapter contains the basic results about the normal fluctuations and the abnormal fluctuations. The difference between normal and abnormal fluctuations depends essentially on the degree of ergodicity of the considered state. This means explicitly the degree of spacial decay to zero at infinity of the space translated truncated correlation functions. A number of properties and applications are also discussed in detail.

We should take note that we discuss only spacial fluctuations. It is of course meaningful and possible to consider as well time-like or dynamical fluctuations. The theory of the time-like fluctuation operators is not yet worked out in full completeness, as its study has been started only recently. It remains an interesting field for future research (see e.g. [87]). It is clear that, in order to get normal time-like fluctuations, the clustering, mixing, or decay properties of the time correlation functions play a crucial role in this case. Also, it is not excluded or it may be expected that again new structural properties may come out for the time fluctuation macroscopic algebra which do not appear in spacial fluctuations.

Another point which must be stressed is that all systems, which are treated in this theory, are quasi-local systems. Boson systems and many other quantum systems, e.g. spin systems, share this property. Other systems, like the full fermion systems, are not quasi-local systems. Their fluctuations need special attention and are not treated here in full detail. Nevertheless fermion sub-systems share some of the properties of quasi-locality. Therefore many of the results which we mention hold true in one or other version also for fermion micro-systems. One of the applications to fermion systems, namely the Luttinger model, is explicitly discussed as an application of the general theory.

In the rest of this section we introduce the basic mathematical structures which are necessary for the description of the concepts and items presented above.

Micro-systems

We introduce first the micro-quantum system. Although all results which we review are straightforwardly extendable to continuous quasi-local systems, sometimes up to some minor technicalities, and therefore also to boson systems, we limit ourselves to quasi-local quantum spin lattice micro-systems. We proceed this way solely for didactic reasons.

We consider the (quasi-)local algebra of micro-observables built on a d-dimensional cubic lattice $\mathbb{Z}^d$. To each lattice point x of the lattice $\mathbb{Z}^d$ we associate an algebra of observables $\mathscr{A}_x$, all copies of the matrix algebra $\mathscr{A} = M_n$, the $n \times n$ complex matrices (n=2 corresponds to spin-(1/2) systems). For all volumes $V \in \mathbb{Z}^d$, the tensor product $\otimes_{x \in V} \mathscr{A}_x$ is denoted by $\mathscr{A}_V$. Every copy $\mathscr{A}_x$ is naturally embedded in $\mathscr{A}_V$. This algebra can be called the algebra of spin observables measurable in the volume V. The family $\{\mathscr{A}_V\}_{V \in \mathbb{Z}^d}$ has the usual relations of *locality* and *isotony* respectively:

$$[\mathscr{A}_{V_1}, \mathscr{A}_{V_2}] = 0 \text{ if } V_1 \cap V_2 = \emptyset \tag{6.2}$$

$$\mathscr{A}_{V_1} \subseteq \mathscr{A}_{V_2} \text{ if } V_1 \subseteq V_2 \tag{6.3}$$

Denote by $\mathscr{A}_L$ the set of all local observables given by

$$\mathscr{A}_L = \cup_V \mathscr{A}_V$$

For clear reasons it is called the *local algebra* and considered as the microscopic algebra of observables of the spin micro-system. Every state ω on this algebra is locally normal, which means that there exists a family of density matrices ρ_V, one for each $V \in \mathbb{Z}^d$, such that

$$\omega(A) = Tr \rho_V A \text{ for each } A \in \mathscr{A}_V$$

The important group of canonical transformations of the local algebra $\mathscr{A}_L$, namely the group of spacial translations $\{\tau_x, x \in \mathbb{Z}^d\}$ defined by

$$\tau_x : A_y \in \mathscr{A}_y \to A_{y+x} \in \mathscr{A}_{y+x},$$

extends straightforwardly to all $A \in \mathscr{A}_L$. We call the local algebra $\mathscr{A}_L$ *asymptotically abelian* for the space translations. This means that for all $A, B \in \mathscr{A}_L$ holds

$$\lim_{|x| \to \infty} [A, \tau_x B] = 0$$

Each observable A commutes with each other, say B, if the latter is translated far enough away in space. This asymptotic abelian property is an immediate consequence of the locality property of the micro-algebra. A state ω on $\mathscr{A}_L$ represents a physical state of the system, assigning to every observable A its expectation value $\omega(A)$. Therefore this setting can be viewed as the quantum analogue of the classical probabilistic setting. Sequences of random variables or observables can be constructed by considering an observable $A \in \mathscr{A}_L$, together with its translates $\tau_x(A)$ for all $x \in \mathbb{Z}^d$, and can be considered as a non-commutative random field.

If a state ω is translation invariant, hence if $\omega \circ \tau_x = \omega$ for all x, then all $\tau_x(A)$ appear as identically distributed random variables. The mixing property of the random field quantum system $(\mathscr{A}_L, \omega)$ is expressed by the spacial correlations tending to zero, that is,

$$\omega(\tau_x(A)\tau_y(B)) - \omega(A)\omega(B) \to 0 \tag{6.4}$$

if $|x - y| \to \infty$. It coincides with what we called before the ergodicity property Eq. (2.26) of the state for boson systems. For what follows, the couple $(\mathscr{A}_L, \omega)$ stands for our physical micro-system. The boson systems $(\mathfrak{A}, \omega)$, considered so far, are as well micro-systems of this type, except for the fact they are not discrete lattice but continuous systems.

Macro-systems

One of the first basic limit theorems of probability theory is the weak law of large numbers. As mentioned already above, in this non-commutative setting the Law of Large Numbers is translated into the problem of the convergence of the space averages of the observables $A \in \mathscr{A}_L$. A first result was given by the mean ergodic theorem of von Neumann (1929). We cited it already (see also [26]) as the following theorem: If the state ω is homogeneous then for all observables A, B and C,

$$\lim_{V \to \mathbb{Z}^d} \omega \left(A \left(\frac{1}{V} \sum_{x \in V} \tau_x B \right) C \right) = \omega(A \lim_V S_V(B)C) \tag{6.5}$$

or equivalently expressed, the sequence of operators

$$S_V(B) = \frac{1}{V} \sum_{x \in V} \tau_x B$$

converges weakly, that is, under the state.

For all ergodic states ω the limit becomes equal to a multiple of the identity operator $S(B) = \lim_V S_V(B) = \omega(B)1$ or formula Eq. (6.5) becomes now:

$$\lim_{V \to \mathbb{Z}^d} \omega \left(A \left(\frac{1}{V} \sum_{x \in V} \tau_x B \right) C \right) = \omega(AC)\omega(B) \tag{6.6}$$

The explicit dependence of the average operator $S(B)$ on the state is clear. This theorem, called the *mean ergodic theorem*, characterizes the class of states yielding a weak law of large numbers. Clearly these limits $\{S(A)|A \in \mathscr{A}_L\}$, which are all multiples of the identity, form a non-trivial but commutative algebra of macroscopic observables. They are called macroscopic because there are infinitely many local operators (all A_x for all x) involved.

Now we go a step further and consider space fluctuations. Let us define for any finite volume V the *local fluctuation* of an observable A in a spacial invariant ergodic state ω by the expression

$$F_V(A) = \frac{1}{V^{1/2}} \sum_{x \in V} (\tau_x A - \omega(A)) \tag{6.7}$$

The following problem is now explicitly posed: How do we give a rigorous mathematical meaning to the quantity $\lim_V F_V(A)$ for V tending to $\mathbb{Z}^d$ in the sense of extending boxes? How do we formulate such a limit? When does such a limit exist? Which are the properties of these quantum fluctuations or of the set of limits $F(A) \equiv \lim_V F_V(A)$? First, it is again clear that the $F(A)$ are macroscopic variables constructed from the micro-system $\mathscr{A}_L$ because they depend on infinitely many strictly local elements A_x for all points x of the lattice.

In order to create an amount of intuitive understanding of what we do, we show the following property: As A and B are strictly local elements (i.e., $A, B \in \mathscr{A}_L$), the commutator expression

$$\sum_{y \in \mathbb{Z}^d} [A, \tau_y B] \in \mathscr{A}_L$$

is again local and an easy computation on the basis of the law of large numbers Eq. (6.6) in the ergodic state ω, yields:

$$
\begin{aligned}
\lim_V [F_V(A), F_V(B)] &= \lim_V \frac{1}{V} \sum_{x \in V} \tau_x \left(\sum_{y \in V} [A, \tau_{y-x} B] \right) \\
&= \lim_V \frac{1}{V} \sum_{x \in V} \tau_x \left(\sum_{y \in \mathbb{Z}^d} [A, \tau_y B] \right) \\
&= \sum_{y \in \mathbb{Z}^d} \omega \left([A, \tau_y B] \right) \equiv i\sigma_\omega(A, B) 1
\end{aligned}
$$

where σ_ω is a bilinear, antisymmetric form on the vector space $\mathscr{A}_L$. Such a form is called a symplectic form. This form σ_ω depends strongly on the state ω. Moreover, if the operators $F(A)$ and $F(B)$ limits do exist as operators acting on one or another Hilbert space, then they satisfy the canonical commutation relations

$$[F(A), F(B)] = i\sigma_\omega(A, B) 1 \tag{6.8}$$

Clearly this result holds for all strictly local elements A and B of the local algebra $\mathscr{A}_L$. As mentioned the right hand side of the equality is a multiple of the unity operator depending on the state ω. This simple property already indicates that fluctuations should not always commute but do have essentially the same commutation relations as the boson fields (see Eq. (2.9)). Earlier work [33, 60] suggested that quantum fluctuations behave like bosons. The problem of the mathematical characterization of these fluctuations remains: Can they be considered as operators, that is, as macroscopic observables? If all that is clear then we can conclude that they satisfy the canonical commutation relations with a very special symplectic form, namely σ_ω, defined on the real vector space underlying the micro-algebra $\mathscr{A}_L$. We should now realize that we are faced with a generalization of the original better known setup of the canonical commutation relations(CCR) developed in Chapter Eq. (2).

Generalized CCR-systems

We are now faced with the introduction of what is called the abstract CCR-algebra built on any arbitrary *symplectic space* (H, σ), where H is a real vector space equipped with a possibly degenerate symplectic form σ. We repeat that σ is a bilinear anti-symmetric form on H.

We now give a short construction of such a CCR-algebra of bosonic observables, which we denote by $W(H, \sigma)$ defined on the symplectic space (H, σ); we also discuss the essentials which we need about the states and representations of this algebra. For more mathematical details we refer to [131], which contains more mathematical properties of this algebra construction. We note that there is little new in [131] compared to what is already explained in Chapter Eq. (2.3), except for the degree of generality and abstractness. In fact, in Chapter Eq. (2.3) the space (H, σ) is identified with the test function space $H = \mathscr{S}$; the symplectic form is given by $\sigma(.,.) = \Im(.,.)$, the imaginary part of the complex scalar product on $L^2(\mathbb{R}^d)$.

Continuing with the generalization, we denote by $W(H, \sigma)$ the complex vector space generated by the linear span of the functions $W(f), f \in H$ defined on H by

$$W(f) : H \to \mathbb{C} : g \to W(f)(g) = \left\{ \begin{matrix} 0 \text{ if } f \neq g \\ 1 \text{ if } f = g \end{matrix} \right\} \tag{6.9}$$

$W(H, \sigma)$ becomes an algebra with the unit $W(0) = 1$ for the product rule

$$W(f)W(g) = W(f + g)e^{-\frac{i}{2}\sigma(f,g)}; f, g \in H$$

We should recognize in this equation the generalized Weyl form of the canonical commutation relations Eq. (2.9). Here, $W(H, \sigma)$ becomes a self-adjoint algebra, also called *-algebra, invariant under the involution defined by

$$W(f) \to W(f)^* = W(-f)$$

For clear reasons the algebra $W(H, \sigma)$ is called the *CCR-Weyl algebra built on the symplectic space* (H, σ).

A linear functional ω on the algebra $W(H, \sigma)$ is again called a state if it is normalized $\omega(1) = 1$, and positive, that is, if $\omega(A^*A) \geq 0$ for all $A \in W(H, \sigma)$. More explicitly, ω is a state on the Weyl algebra $W(H, \sigma)$ if ω is linear, normalized, and positive or for any choice of $A = \sum_j c_j W(f_j)$ holds the linearity and

$$\sum_{j,k} c_j \bar{c}_k \, \omega\left(W(f_j - f_j)\right) e^{-i\sigma(f_j, f_k)} \geq 0$$

$$\omega(1) = \omega(W(0)) = 1.$$

Again every state gives rise to a representation of $W(H, \sigma)$ by means of the GNS-construction (see Eq. (7.1)).

A remark about the special case that σ is possibly *degenerate* is in order. Denote by H_0 the kernel of σ, that is,

$$H_0 = \{f \in H \,|\, \sigma(f, g) = 0 \text{ for all } g \in H\}$$

If H_0 is not trivial, then σ is called *degenerate*. We can write in the direct sum form $H = H_0 \oplus H_1$, where H_1 is the complement of H_0 in H. Let σ_1 be the non-degenerate symplectic form on the subspace H_1. The form σ_1 is the restriction of σ to H_1. We can verify that the algebra $W(H, \sigma)$ is isomorphic to the tensor product of two other Weyl algebras, explicitly

$$W(H, \sigma) = W(H_0, 0) \otimes W(H_1, \sigma_1)$$

Clearly $W(H_0, 0)$ is an abelian algebra and each positive definite normalized functional φ on this subalgebra

$$\varphi : h \in H_0 \rightarrow \varphi(W(h))$$

defines a state $\omega : \omega(W(h)) = \varphi(W(h))$ on $W(H_0, 0)$. Let ξ be any *character* of the additive group H, then the map τ_ξ,

$$\tau_\xi W(f) = \xi(f)W(f)$$

extends to a canonical transformation Eq. (7.3) of the Weyl algebra $W(H, \sigma)$. This transformation is a generalization of the field translation defined in Chapter Eq. (2) and in Eq. (7.3). Let s be a positive symmetric bilinear form on H which is majorizing the symplectic form σ in the following sense: for all $f, g \in H$ holds:

$$\frac{1}{4}|\sigma(f,g)|^2 \le s(f,f)s(g,g) \tag{6.10}$$

Also, let $\omega_{s,\xi}$ be the positive normalized linear functional on $W(H, \sigma)$ given by

$$\omega_{s,\xi}(W(h)) = \xi(h)e^{-\frac{1}{2}s(h,h)} \tag{6.11}$$

It is straightforward [131] to check that $\omega_{s,\xi}$ is a state on the whole algebra $W(H, \sigma)$. All states of this type are called again *quasi-free states* Eq. (2.14) on the CCR-algebra $W(H, \sigma)$ (the generalized Weyl algebra).

A state ω of the algebra $W(H, \sigma)$ is called a *regular state* if, for all $f, g \in H$, the map $\lambda \in \mathbb{R} \rightarrow \omega(W(\lambda f + g))$ is continuous. This regularity property of a state yields (see [131]) the existence of a Bose field as follows: Let $(\mathcal{H}, \pi, \Omega)$ be the GNS-representation Eq. (7.1) of the state ω. The regularity of ω implies that there exists a real linear map $b : H \rightarrow \mathcal{L}(\mathcal{H})$ (linear operators on $\mathcal{H}$) such that $\forall f \in H :$ $b(f)^* = b(f)$ and

$$\pi(W(f)) \equiv W(f) = \exp ib(f)$$

The map b is the *Bose field* satisfying the generalized boson field commutation relations:

$$[b(f), b(g)] = i\sigma(f,g) \tag{6.12}$$

Note that the boson fields are state dependent. However, none of this is terribly new. The boson field discussed in Section Eq. (2.2) is also state dependent. In fact they are the boson fields of the Fock state space representation of the Weyl algebra. Furthermore typical is that at the place where we expect to find Planck's constant, we essentially find an expectation value of a commutator (see Eq. (6.9)). Finally we also note that for all continuous characters ξ of H all corresponding quasi-free states are regular states guaranteeing the existence of boson fields in their representations.

6.2 Normal Quantum Fluctuations

In this section we develop the theory of normal fluctuations for d-dimensional quantum lattice micro-systems. As explained already the latter systems have a quasi-local structure. In order to present the essentials for technical simplicity we assume again that the local micro-algebras $\mathscr{A}_x$, for each point x of the lattice $\mathbb{Z}^d$ are copies of the algebra M_n, the $n \times n$ complex matrices—the typical quantum spin situation. The basic results which are derived below remain valid essentially for all lattice or continuous quasi-local algebras (see [61, 66, 63]).

Hence we consider the *physical micro-system* $(\mathscr{A}_L, \omega)$ where ω is an ergodic state of $\mathscr{A}_L$.

For any local observable A we introduce its local fluctuation operator Eq. (6.1) in the state ω of the micro-system by

$$F_V(A) = \frac{1}{\sqrt{V}} \sum_{x \in V} (\tau_x A - \omega(A)) \qquad (6.13)$$

The problem is to give a rigorous mathematical meaning to the limits

$$\lim_{V \to \infty} F_V(A) \equiv F(A) \qquad (6.14)$$

where the limit is taken for any increasing $\mathbb{Z}^d$-absorbing sequence $\{V\}_V$ of finite volumes V of $\mathbb{Z}^d$ tending to infinity. For simplicity we will consider these volumes as increasing boxes. The limits $F(A)$, once they are established, are called the macroscopic *fluctuation operators* of the micro-system $(\mathscr{A}_L, \omega)$.

As mentioned, early work [33, 60] has already suggested that the fluctuations behave like bosons. These ideas become technically more explicit and complete by proving that we indeed obtain well-defined representations of a generalized CCR-algebras of fluctuations, which are completely determined by the original micro-system $(\mathscr{A}_L, \omega)$.

It is important to keep in mind that we make the following choices: Choose for $H = \mathscr{A}_{L,sa}$ the real vector space of the self-adjoint elements of the micro-algebra $\mathscr{A}_L$ or any of its subspaces. We can formulate the following definitions:

Definition 6.1. *An observable* $A \in H$ *satisfies the normal quantum central limit theorem (CLT) for the ergodic state* ω *if the following limits exist and take the appropriate explicit forms:*

1. $\lim_V \omega(F_V(A)^2) \equiv s_\omega(A, A)$ *exists and is finite, and*

2. $\lim_V \omega(e^{itF_V(A)}) = e^{-(t^2/2)s_\omega(A,A)}$ *for all real numbers* t,

where s_ω *is a positive bilinear symmetric form on the vector space* H.

This definition of central limit theorem coincides with the commutative probabilistic notion in terms of characteristic functions for classical systems, in which case the algebra $\mathscr{A}_L$ is abelian. The classical version refers to the notion of convergence in distribution. For quantum systems we do not have a standard notion of convergence in distribution. Only the concept of convergence in expectations is relevant. On the other hand our definition for the quantum situation does not exclude the notion of the central limit theorem in terms of the moments, which is sometimes called the moment version of the central limit theorem. It amounts to the formulation of the states in terms of the correlation functions.

Definition 6.2. *The micro-system* $(\mathscr{A}_L, \omega)$ *is said to have normal quantum fluctuations for an ergodic ω if*

1. $\forall A, B \in H$

$$\sum_{x \in \mathbb{Z}^d} |\omega(A\tau_x B) - \omega(A)\omega(B)| < \infty$$

2. the CLT holds for all $A \in H$

Note that *1.* implies that the state ω must be ergodic for the space translations. Moreover as a consequence of *1.*, we can define a sesquilinear form on the vector space $\mathscr{A}_L$ by

$$\langle A, B \rangle_\omega = \lim_V \omega\left(F_V(A^*)F_V(B)\right) \tag{6.15}$$
$$= \sum_x \{\omega(A^*\tau_x B) - \omega(A^*)\omega(B)\}$$

and introduce the following notations:

$$s_\omega(A,B) = Re\langle A,B \rangle_\omega \; ; \; \sigma_\omega(A,B) = 2Im\langle A,B \rangle_\omega \tag{6.16}$$

For $A, B \in H$ we obtain

$$\sigma_\omega(A,B) = -i \sum_{x \in \mathbb{Z}^d} \omega([A, \tau_x B]) \; ; \; s_\omega(A,A) = \langle A,A \rangle_\omega \tag{6.17}$$

The couple $(H, \sigma) = (\mathscr{A}_{Lsa}, \sigma_\omega)$ is a symplectic space and s_ω is a non-negative real symmetric bilinear form on the real space H. Following the introductory remarks of the previous section about the generalized algebra of observables, we obtain a natural CCR-algebra $W(H, \sigma_\omega)$ defined on this symplectic space. This algebra of observables depends heavily on the given state ω. The following theorem is an essential step in the construction of a macroscopic physical system of normal fluctuations constructed out of the micro-system $(\mathscr{A}_L, \omega)$:

Theorem 6.3. *If the micro-system* $(\mathscr{A}_L, \omega)$ *has normal fluctuations, then the limits* $\{\lim_V \omega(e^{iF_V(A)}) = \exp\{(-1/2)s_\omega(A,A)\}, A \in H\}$ *define a quasi-free state $\tilde{\omega}$ on the CCR-algebra $W(H, \sigma_\omega)$ by*

$$\tilde{\omega}(W(A)) = \exp\left(-\frac{1}{2}s_\omega(A,A)\right)$$

Proof. The proof is clear from the definitions if we can prove that the positivity condition Eq. (6.10) holds. But the latter follows readily from

$$\frac{1}{4}|\sigma_\omega(A,B)|^2 = \lim_V |Im\,\omega(F_\Lambda(A)F_\Lambda(B))|^2$$
$$\leq \lim_V \omega(F_V(A)^2)\omega(F_V(B)^2) = s_\omega(A,A)s_\omega(B,B).$$

having used the Schwartz inequality.

This theorem indicates once more that the quantum mechanical alternatives for the (classical) Gaussian measures are the quasi-free states on CCR-algebras or boson algebras. However the following question still remains: In the case of normal fluctuations, is it possible to take the limits of products of the form

$$\lim_V \omega \left(e^{iF_V(A)} e^{iF_V(B)} \cdots \right)$$

and if these exist, do they preserve the CCR-Weyl-structure? Clearly this is asking for the typical non-commutative structure of the macro-algebra of fluctuations.

Using the following general bounds for the norm bounded operators $C^* = C$ and $D^* = D$, we derive the inequalities

$$\left\| e^{i(C+D)} - e^{iC} \right\| \leq \|D\|$$

$$\left\| [e^{iC}, e^{iD}] \right\| \leq \|[C,D]\|$$

$$\left\| e^{i(C+D)} - e^{iC} e^{iD} \right\| \leq \frac{1}{2} \|[C,D]\|$$

and by using the expansion of the exponential function we can easily prove the norm limit

$$\lim_V \left\| e^{iF_V(A)} e^{iF_V(B)} - e^{i(F_V(A)+F_V(B))} e^{-\frac{1}{2}[F_V(A),F_V(B)]} \right\| = 0 \qquad (6.18)$$

if A and B are one-point observables (e.g. take $A, B \in \mathscr{A}_{\{0\}}$). For general local elements, the proof is somewhat more technically involved but can be performed on the basis of a Bernstein-like argument (for full details see [63]). The property Eq. (6.18) can be considered as a *Baker-Campbell-Hausdorff formula* for fluctuations. From this formula, the mean ergodic theorem, and the theorem above we indeed obtain the following theorem:

Theorem 6.4. *If the micro-system* $(\mathscr{A}_L, \omega)$ *has normal fluctuations then for $A, B \in H$, for which we can take any subspace of $\mathscr{A}_L$, there exists a quasi-free state $\tilde{\omega}$ on the macro-CCR-algebra $W(H, \sigma_\omega)$ of the quantum fluctuations such that*

$$\lim_V \omega \left(e^{iF_V(A)} e^{iF_V(B)} \right) = \exp \left\{ -\frac{1}{2} s_\omega(A+B, A+B) - \frac{i}{2} \sigma_\omega(A,B) \right\}$$

$$= \tilde{\omega}(W(A)W(B))$$

The two theorems of this section describe completely the topological and analytical aspects of the quantum central limit theorem under the condition of normal fluctuations. We should realize that the quantum central limit yields, for every micro-physical system $(\mathscr{A}_L, \omega)$, many macro-physical boson systems with an algebra of observables $(W(H, \sigma_\omega), \tilde{\omega})$, the CCR-algebra of fluctuation observables in the GNS-representation defined by the quasi-free state $\tilde{\omega}$. Because the state $\tilde{\omega}$ is a quasi-free state, it is also a regular state; the map $\lambda \in \mathbb{R} \to \tilde{\omega}(W(\lambda A + B))$ is therefore continuous for each A and B. From the remarks in the previous section we know that this regularity property yields the existence of a Bose field. in particular there exists a real linear map

$$F : A \in H \subseteq \mathscr{A}_{L,sa} \to F(A)$$

where $F(A)$ is a self-adjoint operator called the field *fluctuation operator*. It is acting on the GNS-representation space $\tilde{\mathscr{H}}$ constructed from $\tilde{\omega}$ (see Eq. (7.1)) such that the following commutation relation holds for all $A, B \in H$:

$$[F(A), F(B)] = i\sigma_{\omega}(A, B).$$

This means that we constructed the *quantum fluctuation boson field* F. It is shown that the quantum central limit procedure realizes the creation of boson systems of fluctuations for all local systems. This result or procedure is sometimes called the operation of *bosonization*.

If we have such a field, we can then ask for their corresponding creation and annihilation operators. One of the main properties of the latter ones is that they are complex linear, respectively complex anti-linear maps defined on the real space (H, σ_{ω}). For this reason we must first make out of this real space a complex space. We consider any *complex structure* on the real symplectic space (H, σ_{ω}), that is, an operator J such that $J^{+} = -J, J^{2} = -1$, where J^{+} is the adjoint of J with respect to the symplectic form (i.e. $\sigma_{\omega}(JA, B) = -\sigma_{\omega}(A, JB)$), which moreover satisfies the property that $\sigma_{\omega}(A, JB) > 0$ for all $A, B \in H$. It is clear that the usual multiplication by the imaginary unit, usually denoted i (i.e., the complex structure of the algebra of observables) is an example of such a complex structure. Why not always select this one? Well, just because it is sometimes physically or formally more rewarding to consider another complex structure. In any case, given a complex structure J, one defines the creation and annihilation operators of the fluctuation boson field F by

$$F^{\pm}(A) = \frac{1}{\sqrt{2}}(F(A) \mp iF(JA))$$

We compute easily that they satisfy the usual boson creation/annihilation commutation relations

$$[F^{-}(A), F^{+}(B)] = \sigma_{\omega}(A, JB) + i\sigma_{\omega}(A, B)$$

These are the *creation and annihilation operators of the quantum fluctuation particles or fields* which we constructed.

Finally, it is straightforward, nevertheless important, to stress once more that all the results of this section about the existence and the properties of a macro-system of boson quantum fluctuations hold true if the linear space of the local micro-observables H is replaced by any of its subspaces. Indeed we should realize that some of these subspaces may be of a greater physical importance than others, a property which depends on the system at hand. This means that the quantum central limit theorems can realize several meaningful macro-physical systems of fluctuations after having started from the same micro-system. In any case, all of them are quasi-free boson field systems. It is also interesting to repeat that these results can be considered as giving a probabilistic basis for the appearance of the quantum canonical commutation relations in general. The essential and basic ingredients are the locality of the micro-system and the central limit theorem.

Coarse Graining

The notion of *coarse graining* is well known in statistical physics as a theory dealing with systems possessing numerous degrees of freedom. It is intrinsically related to the main goals of statistical mechanics, namely to search for a possibly small number of the relevant quantities within the huge number of degrees of freedom of these systems. Now we analyze some aspects of this notion of coarse graining which turn up as a consequence of taking the quantum central limit. Consider again the sesquilinear form $< .,. >_\omega$ Eq. (6.15) on $\mathscr{A}_L$

$$\langle A, B \rangle_\omega = \sum_{x \in \mathbb{Z}^d} (\omega(A^* \tau_x B) - \omega(A)\omega(B)) = s_\omega(A,B) + i\sigma_\omega(A,B) \qquad (6.19)$$

This form presupposes the convergence of the sum in the expression which is itself a necessary condition for the central limit being meaningful. We note that this form defines a topology on the vector space $\mathscr{A}_L$ which can be checked by the more mathematics-minded among us not to be comparable with any of the operator topologies [26] on the GNS-representation induced by the original state ω. In fact, this form topology is not closable in these weak, strong, ultra-weak or ultra-strong operator topologies induced by the state ω. This shows that the quantum central limit is not comparable with these operator topologies.

Now we introduce an equivalence relation on the micro-observables. Call A and B in $\mathscr{A}_L$ *equivalent*, denoted by $A \sim B$, if and only if $\langle A - B, A - B \rangle_\omega = 0$. Clearly this defines an equivalence relation on the vector space $\mathscr{A}_L$ which we can consider as a property of coarse graining, which is mathematically characterized by the following property: For all $A, B \in H = \mathscr{A}_{L,sa}$ the relation $A \sim B$ is equivalent with $F(A) = F(B)$.

Indeed suppose first that $F(A) = F(B)$. Then $[W(A), W(B)] = 0$ and hence $\sigma_\omega(A, B) = 0$. Therefore from the first theorem Eq. (6.3) we obtain

$$1 = \tilde{\omega}(W(A)W(B)^*) = \tilde{\omega}(W(A)W(-B))$$
$$= \tilde{\omega}(W(A-B)) = \exp -\frac{1}{2} s_\omega(A - B, A - B)$$

and hence $\langle A - B, A - B \rangle_\omega = 0$. The converse is as straightforward.

From this property it follows immediately that in particular the action of the translation group acts trivially on the fluctuations, that is, $F(\tau_x A) = F(A)$ for all $x \in \mathbb{Z}^d$. Therefore the map $F : H = \mathscr{A}_{L,sa} \rightarrow W(H, \sigma_\omega)$ is not injective. Although this is a trivial example, it shows that the equivalence relation is not an empty statement. This mathematical property is an expression of the physical phenomenon of coarse graining yielding also a mathematically rigorous formulation of the quantum fluctuations as being genuine macroscopic observables.

Clustering Condition, Scaling Law

Above we construct the new macroscopic physical system of quantum fluctuations for any micro-system with the property of normal fluctuations under the condition that the micro-system satisfies the central limit theorem. The main remaining question is: When does the micro-system exhibit normal fluctuations? In other words, do we know the conditions needed to produce normal fluctuations? We describe a general sufficient clustering (mixing, ergodicity) condition on the micro-state ω in order that the micro-system $(\mathscr{A}_L, \omega)$ exhibits normal fluctuations.

Let V, V' be finite volumes and ω a translation invariant state. Denote

$$\alpha^{\omega}(V, V') = \sup_{\substack{A \in \mathscr{A}_V : \|A\| = 1 \\ B \in \mathscr{A}_{V'} : \|B\| = 1}} |\omega(AB) - \omega(A)\omega(B)|$$

The cluster function $\alpha_N^{\omega}(d)$ is defined by

$$\alpha_N^{\omega}(d) = \sup_{V, V'} \left\{ \alpha^{\omega}(V, V') \ : \ d(V, V') \geq d \ \text{and} \ d(V, V') \leq N \right\}$$

where $N, d \in \mathbb{R}^+$ and $d(V, V')$ is the euclidean distance between the volumes V and V'. Note that we use here the symbol d for the distance and not the dimension of the system. It is straightforward to see that

$$\alpha_N^{\omega}(d) \leq \alpha_N^{\omega}(d') \quad \text{if} \quad d \geq d'$$
$$\alpha_N^{\omega}(d) \leq \alpha_{N'}^{\omega}(d) \quad \text{if} \quad N \leq N'$$

The *clustering condition*, which means the degree of ergodicity of the state ω that we are seeking, is expressed by the scaling law

$$\exists \delta > 0 \ : \ \lim_{N \to \infty} N^{1/2} \alpha_N^{\omega} \left(N^{\frac{1}{2d} - \delta} \right) = 0 \tag{6.20}$$

where d stands again for the dimension. Or equivalently

$$\exists \delta > 0 \ : \ \lim_{N \to \infty} N^{d+\delta} \alpha_{N^2(d+\delta)}^{\omega}(N) = 0 \tag{6.21}$$

This condition implies the somewhat more transparent condition, namely: For each integer N,

$$\sum_{x \in \mathbb{Z}^d} \alpha_N^{\omega}(|x|) < \infty$$

which states that the function $\alpha_N^{\omega}(\cdot)$ is an $L^1(\mathbb{Z}^d)$-function for all N. In fact this condition corresponds to the so-called uniform mixing condition in the commutative (classical) central limit theorem (see [83]). This condition was already used in [74], where the function $\alpha^{\omega}(d)$ is termed the modulus of decoupling.

In the special case of product states, for example the ergodic equilibrium states of mean field boson systems [50], the states are uniformly clustering with $\alpha^{\omega}(d) = 0$ for all $d > 0$. The normality property of the fluctuations of the micro-system for all product states has been proven and extensively studied (see e.g. [61]). The arguments used in this proof, being relatively simple, give nevertheless a good insight in the arguments for the proof of the general theorem. We will now reproduce this proof.

Central Limit Theorem for Product States

Without loss of generality we can limit ourselves in this case to the one-dimensional lattice $\mathbb{Z}$. A *product state* ω on $\mathscr{A}_L$ means a state with the property that, for all $A \in \mathscr{A}_x$ and $B \in \mathscr{A}_y$ with $x \neq y$, the expectation value of the product of two observables at different points of the lattice equals the product of the expectation values, that is, $\omega(AB) = \omega(A)\omega(B)$. Because of this property and the homogeneity of the state, it is uniquely defined by its expectation values at one point on the lattice. Take for this lattice point the singleton point $\{0\}$. The corresponding local algebra is then $\mathscr{A}_0$. As this local algebra is an algebra of matrices, the state ω restricted to $\mathscr{A}_0$ is described by a density matrix, say ρ, and $\omega(A) = tr\,\rho A$ for all local observables A. Such a state is translation invariant, and as a product state is perfectly ergodic.

Following the definition of the quantum fluctuation Eq. (6.7) of the micro-observable A in the product state ω, we can write

$$F_N(A) = \frac{1}{\sqrt{N}} \sum_{x=1}^{N} (\tau_x A - \omega(A)) = \frac{1}{\sqrt{N}} \sum_{x=1}^{N} (\tau_x A - tr\,\rho A)$$

We compute $\lim_N F_N(A)$ in the sense of the central limit theorem; namely we consider $\lim_N \omega(\exp i F_N(A))$. Using the commutators $[\tau_x A, A] = 0$ for all $x \neq 0$, the extreme locality property of A, and the product property of the state (corresponding to the most extreme form of ergodicity) we readily compute

$$\omega(e^{iF_N(A)}) = \omega(\Pi_{x=1}^{N} \exp \frac{i}{\sqrt{N}}(\tau_x A - \omega(A)))$$

$$= \{\omega(\exp \frac{i}{\sqrt{N}}(\tau_x A - \omega(A)))\}^N$$

$$= \{1 - \frac{1}{2N} \omega((A - \omega(A))^2 + \Sigma_{k=3}^{\infty} \frac{i^k}{k!N^{k/2}}(A - \omega(A))^k)\}^N$$

Using that the matrix A is norm-bounded, we obtain

$$|\Sigma_{k=3}^{\infty} \frac{i^k}{k!N^{k/2}} \omega(A - \omega(A))^k| \leq \frac{1}{N^{3/2}} e^{2\|A\|}$$

and therefore in the limit $N \rightarrow \infty$:

$$\lim_N \omega(e^{iF_N(A)}) = e^{-\frac{1}{2}s_\omega(A,A)}$$

where

$$s_\omega(A,A) = \omega(A^2 - \omega(A)^2) = tr\,\rho A^2 - (tr\,\rho A)^2 < \infty$$

This finishes the computation of the relevant formula for product states and the proof of the existence of the central limit and hence the existence of the normal fluctuations for these states.

Quantum Central Limit Theorem

Naturally, the product states share the extreme form of ergodicity and indicate that weaker forms of ergodicity still allow us to obtain the existence of the normal central limit property. The condition Eq. (6.20) is a much weaker but sufficient condition for the central limit property, as proven in [63]. In the latter case, the proofs are rather technical and based on a generalization of the well known Bernstein argument [83] of the classical central limit theorem applied to the non-commutative situation. For the sake of formal self-consistency we formulate the following theorem without proof:

Theorem 6.5. *(Quantum Central Limit Theorem) Take the micro-system* $(\mathscr{A}_L, \omega)$ *such that the state* ω *is lattice translation invariant and satisfies the clustering condition Eq. (6.20). Then the system has normal fluctuations for all elements of the vector space of local observables* $\mathscr{A}_{L,sa}$.

The literature contains several other forms of quantum central limit theorems. In [67] a non-commutative central limit theorem is derived using similar techniques. However the main difference with the theorem Eq. (6.5) is its strictly local character in the sense that the proof is performed for one single local operator separated from the rest of the system. Moreover, the conditions we must satisfy are formulated in terms of the spectral properties of this very operator. The global approach idea from micro to macro resulting in the CCR-algebraic structure of the fluctuations is not at the order and totally absent.

Another non-commutative central limit is obtained in [1] where the method of moments is used, which requires different mixing or clustering conditions. It is not straightforward to check whether a state satisfying these conditions satisfy also the degree of mixing expressed in the conditions Eq. (6.20), or vice versa. However we would expect that all these different conditions, under which the central limit theorems hold, are satisfied for high enough temperature equilibrium states. Cluster expansion techniques may be helpful to prove this. For quantum spin chains, a theorem analogous to the result of the theorem Eq. (6.5) is proven under weaker conditions in [118, 119]. Finally in [150], we find a proof of the quantum central limit theorem for modulated observables and equilibrium states satisfying specific momentum support properties of the correlation functions. We point out that *modulated fluctuation operators* are defined as follows: For any $k \in V^*$, $k \neq 0$ and for any ergodic state ω, we define the *k-mode quantum fluctuation operator* as the central limit $V \rightarrow \infty$ of the local operators

$$F_V(A,k) = \frac{1}{\sqrt{V}} \sum_{x \in V} e^{ikx} (\tau_x A - \omega(A))$$

for any local observable A. These modulated fluctuations reflect closely the results of taking the Fourier transform of the operator, an operation already used for the creation and annihilation operators a_k^* and a_k. Fourier transforms are widely used in the physics literature. Considering k-mode fluctuations or taking Fourier transforms provides a mathematical boson structure to these operations. Now for $k \neq 0$, the k-dependance changes the local character of the fluctuations and a new proof of the central limit must be used. It turns out that the central limit always holds under the scaling-law condition Eq. (6.20), but also that this central limit theorem holds for the modulated fluctuations under much milder conditions. For the details of this proof we refer to [122] and [126]. This finishes the small review about the presence of a variety of different conditions under which the quantum central limit theorem can hold for physical spin micro-systems $(\mathscr{A}_{sa}, \omega)$ having normal fluctuations.

Quantum Fluctuation Dynamics

We extend our attention from the physical system $(\mathscr{A}_L, \omega)$ to the *dynamical system* $(\mathscr{A}_L, \omega, \alpha_t)$ (see Eq. (7.1), [26]), which is the physical system $(\mathscr{A}_L, \omega)$ to which we add a reversible micro-dynamics α_t. We investigate the effect of the micro-dynamics α_t on the fluctuations macro-CCR-algebra $W(H, \sigma_\omega)$. In other words, we examine the dynamics after having taken the central limit. As usual the micro-dynamics is supposed to be of the short-range type to guarantee the existence of this dynamics as a norm limit of the usual type

$$\alpha_t(\cdot) = \lim_V e^{itH_V} \cdot e^{-itH_V}$$

and we assume it to be space translation invariant that is $\alpha_t \cdot \tau_x = \tau_x \cdot \alpha_t$, $\forall t \in \mathbb{R}$, $\forall x \in \mathbb{Z}^d$. We suppose that the state ω is as well space as time translation invariant, that is for all x: $\omega \cdot \tau_x = \omega$ and for all t: $\omega \cdot \alpha_t = \omega$. Moreover we assume that the state ω satisfies the sufficient condition Eq. (6.20) for normal fluctuations.

For every local observable $A \in \mathscr{A}_{L,sa}$ we defined the local fluctuation $F_V(A)$ and obtained a clear meaning for the limit $F(A) = \lim_V F_V(A)$ from the central limit theorem. Now we are interested in the macro-dynamics of the fluctuations $F(A)$ induced by the micro-dynamics (α_t). Clearly for all A and all finite volumes V,

$$\alpha_t F_V(A) = F_V(\alpha_t A)$$

Therefore we are tempted to define the macro-dynamics $\tilde{\alpha}_t$ or the dynamics of the fluctuations in the volume limit by

$$\tilde{\alpha}_t F(A) = F(\alpha_t A) \tag{6.22}$$

Note however that the locality of the micro-observable is an important property in order to prove the existence of its fluctuation operator, and that in general $\alpha_t A$ is no longer a local element of $\mathscr{A}_{L,sa}$. It is a priori unclear whether the central limit for elements of the type $\alpha_t A$, with $A \in \mathscr{A}_{Lsa}$ exist or not, and hence whether we can ascribe a meaning to the expression $F(\alpha_t A)$. Moreover if $F(\alpha_t A)$ exists, it remains to show that the set of maps $(\tilde{\alpha}_t)_t$ in Eq. (6.22) again defines a continuous one-parameter group of canonical transformations Eq. (7.3). The latter ones are now acting on the fluctuation macro-CCR-algebra. More precisely they act on the $\tilde{\omega}$-weak closure $W(\mathscr{A}_{L,sa}, \sigma_\omega)''$ of the macro-algebra $W(\mathscr{A}_{L,sa}, \sigma_\omega)$ (This is also the von Neumann algebra generated by the $\tilde{\omega}$-GNS-representation of $W(\mathscr{A}_{Lsa}, \sigma_\omega)$). For the definition of $\tilde{\omega}$ see Theorem Eq. (6.3) Eq. (6.4). In other words, we have to show that the bona fide macro-dynamics $(\tilde{\alpha}_t)_t$ really exists. All of the mentioned steps need a proof. Ref. [63] proves the following basic theorem concerning the existence of these dynamics:

Theorem 6.6. *Under the conditions on the dynamics α_t and on the state ω expressed above, the limit $F(\alpha_t A) = \lim_V F_V(\alpha_t A)$ exists in the central limit sense, and the maps $\tilde{\alpha}_t$ extend to a weakly continuous one-parameter group of canonical transformations of the $\tilde{\omega}$-weak closure of the macro-algebra $W(\mathscr{A}_{Lsa}, \sigma_\omega)$. The quasi-free macro-state $\tilde{\omega}$ (Eq. (6.3), Eq. (6.4), Eq. (6.5)) of the fluctuations is either $\tilde{\alpha}_t$-invariant also expressed as macro-dynamics time invariant.*

This theorem yields indeed the existence of a dynamics, $\tilde{\alpha}_t$, on the fluctuations macro-algebra, induced by the micro-dynamics and shows that it is also of the quasi-free or the solvable type. The fluctuation fields evolve in time following the same properties as the solvable boson dynamics (see Chapter Eq. (5)), which is explicitly expressed by

$$\tilde{\alpha}_t F(A) = F(\alpha_t A)$$

where $F(A)$ is a representation of a boson field in a quasi-free state $\tilde{\omega}$. In physical terms this means that any micro-dynamic α_t always induces a linear or solvable process on the level of its fluctuations.

Altogether, we conclude that indeed the quantum central limit theorem realizes a map from the micro-dynamical system $(\mathscr{A}_{L,sa}, \omega, \alpha_t)$ to the macro-dynamical system $(W(\mathscr{A}_{L,sa}, \sigma_\omega)'', \tilde{\omega}, \tilde{\alpha}_t)$ of the quantum fluctuations. The macro-system is a quasi-free boson system in the sense that the macro-state $\tilde{\omega}$ is a quasi-free state and that the macro-dynamics system is solvable.

It is interesting to note that, as the central limit theorem, also the law of large numbers maps local micro-observables into macroscopic observables, namely by taking the averages. However the latter form a commutative algebra of macro-observables. The law of large numbers also maps the micro-dynamical system into a macro-dynamical systems. But the micro-dynamics is, contrary to the central limit situation, mapped into a trivial macro-dynamics. These considerations make clear that on the level of the law of large numbers we do not expect to observe genuine quantum phenomena. On the other hand, on the level of the fluctuations we have observed that macroscopic quantum phenomena are detectable in nature.

Apart from considering the effects of reversible dynamics on the fluctuations, it is of course also possible to study the effects of the irreversible micro-dynamics. In this case the micro-dynamics is mapped into a solvable irreversible dynamics for the macro-system of fluctuations. For more details about this topic see [62, 64].

6.3 Abnormal Quantum Fluctuations

The results about normal fluctuations contain two essential elements. On the one hand the central limit should hold. A sufficient condition in order that this exists is the validity of the cluster condition Eq. (6.20). The latter condition is on the micro-state ω guaranteeing the normality of the fluctuations. On the other hand there is the reconstruction theorem, identifying the CCR-algebra representation of the fluctuation operators in the emerging quasi-free macro-state we denoted by $\tilde{\omega}$.

The cluster (mixing) condition is in general not satisfied for systems with long-range correlations, a situation showing up for instance at phase transitions in equilibrium states at low temperatures. This situation is also present for instance in boson systems when there is boson condensation. It is a challenging question to study also in this case the problem of existence of fluctuations operators, and if they exist, to study their mathematical structure. We might expect to detect other structures different from the CCR-structure and other types of macro-states which are not quasi-free states.

Progress in the elucidation of all these questions started with a detailed study of abnormal fluctuations in the harmonic and anharmonic crystal models [164, 27, 127]. We obtain other more general Lie algebras than the Heisenberg Lie algebra of the CCR-algebra and more general quantum states $\tilde{\omega}$ which are not quasi-free. In the case of abnormal fluctuations, quantum states are indeed computed which reach far outside the set of quasi-free states that appear when dealing with normal fluctuations.

Abnormal fluctuations turn up if we have an ergodic micro-state ω showing spacial long range correlations. We have in mind continuous (second order) phase transitions. In these situations it is typical that, for instance, the heat capacity or other susceptibilities diverge at critical points, lines, or planes. This means that normally scaled (with the factor $V^{-1/2}$) fluctuations of some specific observables diverge. This statement is detectable in the divergence of sums of the type

$$\sum_{x \in \mathbb{Z}^d} (\omega(A\tau_x A) - \omega(A)^2)$$

for one or more local observable A.

In order to deal with these situations, one way of proceeding is to rescale again the volume of the local fluctuations. What do we do? We introduce a *scaling index* δ_A, a real number in the interval $(-1/2, 1/2)$ that depends in general on the observable A under consideration, and such that the *abnormally scaled* local fluctuations

$$F_V^{\delta_A} = V^{-\delta_A} F_V(A) = \frac{1}{V^{(1/2+\delta_A)}} \sum_{x \in V} (\tau_x A - \omega(A))$$

have a nontrivial characteristic function $\phi_A(.)$ defined by the following limit formula: $\forall t \in \mathbb{R}$,

$$\lim_V \omega_V (e^{it F_V^{\delta_A}(A)}) \equiv \phi_A(t) \qquad (6.23)$$

We limit our discussion to locally normal micro-states ω such that their local restrictions ω_V are normal local Gibbs states.

It is clear that the index δ_A is a measure for the degree of *abnormality of the fluctuation* operator of the observable A in the state ω. Note that $\delta_A = -1/2$ would yield a triviality and that $\delta_A = 1/2$ would lead to the law of large numbers (theory of averages). We should observe here that in this general case the characteristic functions ϕ_A (or the corresponding macro-states $\tilde{\omega}$ of the fluctuations) need not always be Gaussian or quasi-free.

The physics literature, see for example [19], usually describes the long-range order by means of the asymptotic form of the connected or truncated two-point function in terms of the so-called *critical exponent* η_A which is defined in the formula

$$\omega_V(A\tau_x A) - \omega_V(A)^2 \simeq 0\left(\frac{1}{|x|^{d-2+\eta_A}}\right) \text{ for } |x| \to \infty$$

The scaling index δ_A (Eq. (6.23)) is related to the critical exponent η_A by the straightforward relation $\eta_A = 2 - 2d\,\delta_A$.

As stated in Eq. (6.23), the index δ_A is determined by the existence of the central limit. It is explicitly computable in model calculations. See for example the computations in [164, 10, 11, 98, 168, 32] performed for equilibrium states of a number of different models. Apart from their very model dependence, the indices also depend greatly on the chosen boundary conditions. This fact may draw a light on the so-called universality property of the critical exponents, which is different from the existing one.

Suppose now that the indices δ_A are determined by the existence of the central limit Eq. (6.23). The next problem is to find out whether in these cases a reconstruction theorem, comparable to the normal fluctuations case, can also be established giving once again a mathematical sense to the limits

$$\lim_V F_V^{\delta_A}(A) \equiv F^{\delta_A}(A) \tag{6.24}$$

as operators, in general unbounded, acting on a Hilbert space. At this point we are not addressing all the mathematical details of these questions, which are however answered positively in the literature. On the other hand we discuss a proof showing the emergence of the Lie-algebra character of abnormal fluctuations under the following clustering conditions: Condition A: the δ-indices are determined by the existence of the variances (second moments), and Condition B: the third moments exist. Below we formulate these conditions more explicitly. For more details we refer to the reference [127].

We remind ourselves that any algebra $\mathcal{G}$, with an n-dimensional underlying vector space with a basis $\{v_i\}_{i=1,\dots n}$ If $\mathcal{G}$ is equipped with a Lie-product

$$v_j \cdot v_k \equiv [v_j, v_k] = \sum_{l=1}^{n} c_{jk}^l v_l \tag{6.25}$$

with structure constants (c_{jk}^l) satisfying the usual properties

$$c_{jk}^l + c_{kj}^l = 0$$

$$\sum_r (c_{ij}^r c_{rk}^s + c_{jk}^r c_{ri}^s + c_{ki}^r c_{rj}^s) = 0$$

then $\mathcal{G}$ is called a *Lie-algebra*.

Here it is the idea of taking for the local micro-algebra per lattice point an a priori Lie-algebra. In order to fix idea's, for spin systems, we can think of the Lie-algebra of the Lie-group $SU(n)$. In general we consider the concrete Lie algebra basis of operators in the strict local algebra of observables $\mathscr{A}_0$ namely $\{L_0 = i1, L_1, \ldots, L_m\}$, $m < \infty$ such that $L_j^* = -L_j$, $j = 1, 2, \ldots m$ and $\omega(L_j) = \lim_V \omega_V(L_j) = 0$ for $j > 0$. Clearly $\omega_V(L_0) = i$ for all V and the $\{L_i\}$ satisfy the equations Eq. (6.25). Because of the special choice of L_0, we have $c_{ok}^l = c_{ko}^l = 0$ and $c_{jk}^o = -i \lim_V \omega_V([L_j, L_k])$.

We consider now the fluctuations of these generators and look for a characterization of the Lie algebra of these fluctuation operators.

For a state ω with local restrictions ω_V, such that the limit $\omega = \lim_V \omega_V$ is ergodic, consider first the local fluctuations: For $j = 1, \ldots m$

$$F_{j,V}^{\delta_j} = \frac{1}{V^{1/2 + \delta_j}} \sum_{x \in V} (\tau_x L_j - \omega_V(L_j)) \tag{6.26}$$

and for notational convenience set the first one equal to $F_{0,V} = i1$. For completeness purposes we formulate now the above conditions A and B explicitly for this concrete Lie-algebra case:

Condition A: We assume that the parameters δ_j are determined by the existence of the finite and non trivial variances: For all $j = 1, \ldots m$

$$0 < \lim_V \omega_V \left((F_{j,V}^{\delta_j})^2 \right) < \infty \tag{6.27}$$

After reordering the indices, we establish that $1/2 > \delta_1 \geq \delta_2 \geq \cdots \geq \delta_m > -1/2$. Condition B: Assume that all third moments are finite, that is,

$$\lim_V \left| \omega_V \left(F_{j,V}^{\delta_j} F_{k,V}^{\delta_k} F_{l,V}^{\delta_l} \right) \right| < \infty$$

for all j, k, l.

We have in mind that the ω_V are Gibbs states for some local Hamiltonians with some specific boundary conditions. The limit as V tends to infinity may depend exceedingly on these boundary conditions in the sense that they are visible in the values of the indices δ_j (see e.g. [164]). If for some $j \geq 1$, it happens that the corresponding $\delta_j = 0$, then the operator L_j has a normal fluctuation operator. In general we define the fluctuation operators of the generators (L_i) by

$$F_j^{\delta_j} = \lim_V F_{j,V}^{\delta_j} \tag{6.28}$$

where the limit is understood in the sense of Condition A, namely with a finite nontrivial variance. If for some $j \geq 1$, the corresponding $\delta_j \neq 0$, then the fluctuation is called an *abnormal fluctuation operator*. In order to satisfy Condition A, it happens sometimes that we must choose δ_j as negative (see e.g. [164]). As already mentioned, in order to exclude trivial situations it is reasonable to limit the discussion to the case that all $\delta_j > -1/2$.

In any case, on the basis of Condition A, the limit set $\{F_j^{\delta_j}\}_{j=0,\dots m}$ of fluctuation operators generates a Hilbert space $\mathscr{H}$ with scalar product

$$\left(F_j^{\delta_j}, F_k^{\delta_k}\right) = \lim_V \omega_V \left((F_{j,V}^{\delta_j})^* F_{k,V}^{\delta_k}\right) \tag{6.29}$$

On the basis of Condition B, the fluctuation operators are defined as multiplication operators of the Hilbert space $\mathscr{H}$. Note that the Conditions A and B are in general insufficient conditions to obtain a limit characteristic function. However they are sufficient to obtain the notion of a fluctuation operator. Now we proceed with the question of the Lie algebra character of these fluctuation operators acting on the Hilbert space $\mathscr{H}$.

Consider the Lie product of two local fluctuations for a finite volume V. From Eq. (6.25),

$$\left[F_{j,V}^{\delta_j}, F_{k,V}^{\delta_k}\right] = \sum_{l=0}^m c_{jk}^l(V) F_{l,V}^{\delta_l} \tag{6.30}$$

with volume dependent coefficients

$$c_{jk}^l(V) = \frac{c_{jk}^l}{V^{1/2+\delta_j+\delta_k-\delta_l}} \; ; \; l = 1, \dots m$$

$$c_{jk}^0(V) = V^{-\delta_j-\delta_k} \sum_{l=0}^m c_{jk}^l \omega_V(F_{l,V}^{\delta_V})$$

It is a straightforward exercise to check that the $\{c_{jk}^l(V)\}$ are the structure coefficients of a Lie algebra which we denote by $\mathscr{G}(V)$. Hence by considering local fluctuations, we can construct for each volume V a map from the Lie algebra $\mathscr{G}$ onto the Lie algebra $\mathscr{G}(V)$ by a non-trivial explicitly given transformation of the original structure constants. When the transformed structure constants approach their well-defined limit, a new non-isomorphic Lie algebra might appear. The limit algebra $\mathscr{G}(\mathbb{Z}^d) = \lim_V \mathscr{G}(V)$, called the contracted Lie algebra of the original algebra $\mathscr{G}$, is always non-semi-simple. This contraction is a typical *Inönü-Wigner contraction* [85, 86]. The limit algebra $\mathscr{G}(\mathbb{Z}^d)$ is straightforwardly obtained by taking the limit $V \to \infty$ of the above volume-dependent structure constants yielding the final structure constants results (see also [127]):

$$\lim_V c_{jk}^l(V) = \begin{cases} 0 & \text{if } \frac{1}{2} + \delta_j + \delta_k - \delta_l > 0 \\ c_{jk}^l & \text{if } \dots\dots\dots\dots = 0 \\ 0 & \text{if } \dots\dots\dots\dots < 0 \end{cases} \tag{6.31}$$

It is interesting to analyze the possible Lie algebra limits and to distinguish the following special cases:

1. If all fluctuations are normal, we recover the Heisenberg Lie algebra of the canonical commutation relations with the right symplectic form σ_ω of the normal fluctuations algebra, that is, we have the bosonization phenomenon.

2. If $1/2 + \delta_j + \delta_k - \delta_\ell > 0$ for all j, k, l, we obtain an abelian Lie algebra of fluctuations. The corresponding macro-system is a commutative or classical system.

3. We obtain the richest structure if $1/2 + \delta_j + \delta_k - \delta_l = 0$ for all j, k, l, or for some of the indices. In this case we uncover a phenomenon of volume scale invariance, in the sense that the constants $c^l_{jk}(V)$ are V-independent. Algebras different from the CCR-algebra are observed. A particularly interesting case shows up if $\delta_j = -\delta_k \neq 0$. In this case at least one of the indices is negative. Consider the case that $\delta_j < 0$. The corresponding fluctuation $F_j^{\delta_j}$ shows a property of *space squeezing*. The other parameter of the pair is then strictly positive, namely $\delta_k > 0$. The corresponding fluctuation $F_k^{\delta_k}$ shows the property of *space dilatation*. These phenomena are observed and explicitly computed in several models (see e.g. [164]). In particular we can conclude that this analysis yields a microscopic explanation of the macro-phenomenon of squeezing (squeezed states and all that) experimentally observed mainly in the field of quantum optics. In the applications below we shall meet this phenomenon as one of the basic properties in the explicit construction of the Goldstone normal modes of the Goldstone boson particle which can appear in short range interacting systems. It turns out to occur as an intrinsic property of the phenomenon of spontaneous symmetry breakdown.

6.4 Applications

The notion of fluctuation operator as presented above, and the mathematical structure of the algebra of fluctuations have been tested in several solvable models. Many applications of this theory of quantum fluctuations can be found in the list of references. We do not enter into the details of everything in this list, but we limit ourselves to mentioning a number of applications which are enough of a general nature or which are showing typical model independent features highlighting the universal character of the theory. We discuss first the Luttinger model because, although it is a non-solvable system model, we can compute rigorously its fluctuation dynamical macro-dynamical system. This model is also chosen to illustrate the bosonization procedure. We explain its so-called "exact solution" as an application of the quasi-freeness property of the macro-systems of the normal density fluctuations.

6.4.1 Luttinger Model

The literature about the Luttinger model [111] is enormous. Originally the model was intended to be a prototype of a one-dimensional interacting fermion micro-system realistic enough and still solvable. The interest in this model got a revival due to the invention of the "Luttinger-liquid" behavior of normal metals [5].

We discuss this fermion micro-model as a prototype example of the emergence of a meaningful boson solvable macro-system (and hence boson liquid) features, which we show to be a consequence of the central limit theorem and of the notion of quantum fluctuations operators. It is surprising that in the past so little attention has been paid to this aspect because already in [103] it is explicitly stated that the infinite volume limit of this fermion system reveals a new physical behavior. It showed boson aspects quite absent in the finite volume situation. In [165] a careful analysis is given of the infinite volume formulation of the model posing explicitly the problem of the structural emerging of these completely new quasi-particles. We describe the infinite volume limit of the so-called Lieb-Mattis exact solution of this non-solvable Luttinger fermion micro-system model. It turns out that it earns his place in the frame of our bosonization procedure explained above. Moreover the study clearly illustrates that this exact solution should be considered as the exactness of the random phase approximation for this model or as the exact solvability of the collective excitations (zero sound) equation as it is formulated in the microscopic theory of the normal Fermi liquids [133]. As far as we are concerned with the understanding of this solution in the frame of a general many-body theory, the so-called exactness of the solution coincides with the quasi-freeness or the solvability of the macroscopic dynamical system of the normal quantum fluctuations of the model.

Microscopic Luttinger Model

The micro-system comprises two types of fermions indexed by the indices $i = 1$ and $i = 2$. The algebra of micro-observables $\mathscr{A}$ is the algebra generated by the fermion creation and annihilation operators $a_i^*(f)$ and $a_i(f)$, $f \in \mathscr{S} \subset L^2(\mathbb{R})$ where

$$a_i^*(f) = \int dx\, f(x) a_i^*(x) = \int dk\, \widehat{f}(k) a_i^*(k)$$

with $\widehat{f}$ the Fourier transform of f. The operators $a_i(f)$ and $a_i^*(f)$ are acting on the fermion Fock space $\mathfrak{F}_{aF}$. This space is fermion analogue of the boson Fock space. It is built on the fermion vacuum vector Ω_{aF}, uniquely determined by the property that it is annihilated by all annihilation operators of the micro-system, that is, $a_i(f)\Omega_{aF} = 0$ for all f. The fermion Fock space $\mathfrak{F}_{aF}$ is generated by the set of vectors of the type

$$a_{i_1}^*(f_1)...a_{i_n}^*(f_n)\Omega_{aF}$$

for all n and test functions f_i. The operators $a_i(x)^*$ and $a_i(y)$ satisfy the usual *canonical anti-commutation relations (CAR)*

$$\{a_i(x), a_j^*(y)\} \equiv a_i(x)a_j^*(y) + a_j^*(y)a_i(x)$$
$$= \delta_{i,j}\, \delta(x-y) \text{ and } \{a_i(x), a_j(y)\} = 0 \qquad (6.32)$$

The microscopic dynamics of the model is again given by the local Hamiltonians H_L which are the sum of two terms $H_L = H_L^0 + H_L^1$ for each value L (the length of the interval $[L/2, -L/2] \subset \mathbb{R}$) and where H_L^0 is called the free Hamiltonian of the special form

$$H_L^0 = \int_{-L/2}^{L/2} dx \{a_1^*(x)\frac{1}{i}\nabla a_1(x) - a_2^*(x)\frac{1}{i}\nabla a_2(x)\} \qquad (6.33)$$

It is called the free particle Hamiltonian because it is quadratic in the creation and annihilation operators. The local Hamiltonian H_L^1 is the interaction part of the Hamiltonian

$$H_L^1 = 2\lambda \int_{-L/2}^{L/2} dx \int_{-L/2}^{L/2} dy\, a_1^*(x) a_1(x) V(x-y) a_2^*(y) a_2(y) \qquad (6.34)$$

where $V(x) = \frac{1}{L}\sum_{q \in L^*} v(q) e^{-iqx}$ is an even, real function with a bounded Fourier transform $v(k) = \int dx\, e^{ikx} V(x)$, satisfying for all $k \in L^* = \{(2\pi/L)\mathbb{Z}\}$ the conditions

$$(i)\, v(k=0) = 0,\ (ii)\, |\lambda v(k)| < \pi \text{ and } (iii) \sum_k |k| |v(k)|^2 < \infty \qquad (6.35)$$

The Hamiltonian is essentially self-adjoint on a dense domain of vectors of the fermion Fock space and is not bounded from below. For this last reason the model is not optimal for the description of a normal, stable, thermodynamical many-body system.

A number of supplementary but relevant remarks can be formulated. First, the model is not solvable if we define the solvability of a fermion systems analogously with the definition we used for boson systems (see Eq. (3.13)). Computing the expectation the Hamiltonian density in any homogeneous state, we indeed obtain a non-trivial presence of four-point functions due to the presence of the interaction terms.

Further, we intend to apply the theory of fluctuations, but the fermion algebra is not a local algebra because of the appearance of anti-commutation relations instead of the commutation relations. Nevertheless we want to continue with the description and rigorously understanding the position of the so-called "exact solution" given by Lieb and Mattis [103]. It is anyhow clear that there is some work to do in order to prepare the system ready for the application of the fluctuation theory. To deal with all these points, it is instructive to consider first the free, or the non-interacting, Luttinger model before we start with the analysis of the full interacting model.

Micro/macro-dynamics for the Free Luttinger Model

Using the free Hamiltonian Eq. (6.33) in the thermodynamic limit $L \to \infty$, the resulting dynamics, as defined in Chapter Eq. (5), is easily computed and yields the free time evolution α_t^0 given by

$$\alpha_t^0(a_1^*(f)) = a_1^*(e^{itP}f) \text{ and } \alpha_t^0(a_2^*(f)) = a_2^*(e^{-itP}f) \qquad (6.36)$$

where $P = \frac{1}{i}\nabla$, is the usual one-particle quantum mechanical momentum operator. Clearly the dynamics α_t^0 is of the same type as the free particle dynamics (analogous as discussed in Chapter Eq. (5)), however with a one-particle kinetic energy spectrum being linear in the momentum operator instead of quadratic.

We look first for the ground state of the finite-volume free system with the Hamiltonian Eq. (6.33) expressed in Fourier form: For all $k \in L^*$, with $a_i(k) = \frac{1}{\sqrt{L}}\int dx\, e^{ikx} a_i(x)$, the free Hamiltonian becomes

$$H_L^0 = \sum_k k(a_1^*(k)a_1(k) - a_2^*(k)a_2(k)) \tag{6.37}$$

Let us denote by $\omega_{0,L}$ the finite-volume ground state (see Eq. (3.7) with $\beta \to \infty$) for this free system. For notational commodity we consider only the case of zero chemical potential; the case of arbitrary values μ of the chemical potential is treated analogously, because this case is obtained by a simple shift of the momentum. The ground state conditions for the finite volume(L) ground state $\omega_{0,L}$, are explicitly given by the inequalities

$$\omega_{0,L}(X^*[H_L^0, X]) \geq 0$$

for each obserable X. In particular, after the substitutions $X = a_i(k)$; $i = 1, 2$, we obtain the inequalities

$$-k \omega_{0,L}(a_1^*(k)a_1(k)) \geq 0$$
$$k \omega_{0,L}(a_2^*(k)a_2(k)) \geq 0$$

yielding that the state $\omega_{0,L}$ has the Fock vacuum property for the 1-particle (particle with index 1) if $k > 0$, and for the 2-particle if $k < 0$. Substituting $X = a_i^*(k)$ in the ground state definition yields the anti-Fock or the fermi sea property for the 1-particle, if $k < 0$, and for the 2-particle, if $k > 0$. By setting $X = a_1^{(*)}(k) + a_2^{(*)}(k')$ we obtain the product property

$$\omega_{0,L}(a_1^{(*)}(k))\omega_{0,L}(a_2^{(*)}(k')) = \omega_{0,L}(a_1^{(*)}(k)a_2^{(*)}(k')) = 0$$

for the ground state. All of these properties together determine the unique ground state $\omega_{0,L}$, which, again on the basis of these properties, can be called shortly the Fock-anti-Fock(FaF) state. The connotation "FaF" is reasonable because the ground state for each of the particles is half empty and half full. Let us now consider the GNS-representation Eq. (7.1) of this ground state with representation Hilbert space denoted by $\mathfrak{F}_{FaF}$. We omit the reference to the volume V. Similarly denote the cyclic vector by Ω_{FaF}. We obtain $\omega_{0,L}(X) = (\Omega_{FaF}, X \Omega_{FaF})$.

Continuing for finite volume L, we define the following density operators for all $p > 0$:

$$\rho_{1,L}(p) = \sum_k a_1^*(k+p)a_1(k); \qquad \rho_{1,L}(-p) = \sum_k a_1^*(k)a_1(k+p) \tag{6.38}$$

$$\rho_{2,L}(p) = \sum_k a_2^*(k+p)a_2(k); \qquad \rho_{2,L}(-p) = \sum_k a_2^*(k)a_2(k+p) \tag{6.39}$$

A straightforward and easy computation yields the following non-trivial commutation relations of these density operators acting on the Hilbert space $\mathfrak{F}_{FaF}$:

$$[\rho_{1,L}(-p), \rho_{1,L}(p')] = \frac{Lp}{2\pi}\delta_{p,p'} \tag{6.40}$$

$$[\rho_{2,L}(p), \rho_{2,L}(-p')] = \frac{Lp}{2\pi}\delta_{p,p'} \tag{6.41}$$

If we were so inclined, we could verify that these operators do commute if they would be considered as acting on the original fermion Fock space $\mathfrak{F}_{aF}$. These different results are a consequence of comparing two different representations of the CAR-algebra of observables, one with cyclic vector Ω_{aF} and the other one with cyclic vector Ω_{FaF}. The representation of the fermion observables as acting on the FaF-state space is in fact also implicitly used by the authors of the work [103] to obtain their bosonization phenomenon in the Luttinger model.

We remarked already that the CAR-algebra is not (quasi-)local. To circumvent this obstacle we restrict our system by replacing the full CAR-algebra $\mathscr{A}$ as the algebra of observables using the so-called *even CAR-algebra* $\mathscr{A}_e$ which is the subalgebra of the full CAE-algebra generated by the even products of creation and annihilation operators. This restricted algebra of observables has achieved wide acceptance among physicists because it contains at least the gauge invariant fermion observables. But there is more, for this even-algebra $\mathscr{A}_e$ has the agreeable property of being (quasi-)local, which is an overall a priori condition to be able to apply our analysis of the quantum fluctuations. Naturally by definition, the constructed ground state $\omega_{0,L}$ remains a ground state if it is restricted to this even-algebra. Furthermore the thermodynamic limit $\omega_0 = \lim_L \omega_{0;L}$ exists. It is readily obtained and completely determined by the two-point function

$$\omega_0(a_j^*(f_x)a_{j'}(f)) = \delta_{j,j'}\{\int_0^\infty dk e^{ikx}|\widehat{f}(k)|^2\delta_{j,2} + \int_{\infty}^0 dk e^{ikx}|\widehat{f}(k)|^2\delta_{j,1}\} \qquad (6.42)$$

and the vanishing one-point function

$$\omega_0(a_j(f)) = 0; \ j = 1, 2$$

By the way, this ground state, determined by its one- and two-point functions belongs to the set of fermion quasi-free states. These are the fermion analogues of the quasi-free states described before for bosons Eq. (2.14). In the fermion case, a subset of these quasi-free states, namely those with a fixed number of particles, are better known in physics as the *Slater-determinant states*.

Again f_x we denote the function f translated over the distance x. From the expression Eq. (6.42) we can conclude that the limit for x tending to infinity approaches zero faster than any polynomial for all f of compact support. Therefore the state ω_0 certainly satisfies the mixing (scaling) property Eq. (6.20). So far, the free dynamical micro-system $(\mathscr{A}_e, \omega_0, \alpha_t^0)$ is completely prepared as a micro-system satisfying all sufficient conditions for a straightforward application of the theory of normal fluctuations. In particular the central limit theorem for normal fluctuations can be applied immediately. In this application to the free Luttinger model, we can compute the variances in the FaF-state ω_0 of all the relevant local micro-operators. In fact in this application we restrict the micro-algebra to the algebra generated by the particle densities which we defined in Eq. (6.38) and Eq. (6.39).

In view of this computation we must compute the variances for the number operators. We note that the expectation values $\omega_0(a_i^*(x)a_i(x))$ for all positions x are undefined because the *FaF*-state contains infinitely many particles of both types. For this reason, if we want to compute the density fluctuations, we must be a bit careful and compute first the fluctuations of the strictly local operators $N_i(f) = a_i^*(f)a_i(f)$, where f is any test-function with a finite local support. At the end of this computation, we take for f a δ-function convergent sequence of functions, say at a point $x = 0$, or in Fourier form a sequence of functions $\widehat{f}_n$ tending to the constant function $1/\sqrt{2\pi}$. Therefore in what follows, if we use the thermodynamic limit notation $L \to \infty$, we mean always the double limit, first $L \to \infty$ and then $\widehat{f} \to 1/\sqrt{2\pi}$. For any $q \neq 0$, let us consider the q-density fluctuation operators, with $i = 1, 2$,

$$F_L^q(N_i) = \sqrt{\frac{2\pi}{|q|}} \frac{1}{\sqrt{L}} \int_{-L/2}^{L/2} dx \, e^{iqx} (\tau_x N_i(f) - \omega_0(N_i(f)))$$

where we used the notation $N_i \equiv N_i(f)$ which, in view of the remark before, should not create any confusion. We find, for $q, q' \neq 0$,

$$\lim_L \omega_0(F_L^q(N_1)F_L^{q'}(N_1)) = \delta_{q,-q'}\theta(-q) \tag{6.43}$$

$$\lim_L \omega_0(F_L^q(N_2)F_L^{q'}(N_2)) = \delta_{q,-q'}\theta(q) \tag{6.44}$$

where θ is the well known θ-function ($\theta(x) = 1$ if $x \geq 0$, and $\theta(x) = 0$ if $x < 0$). We compute as well in the ω_0-weak sense (i.e., under the state ω_0) the commutators

$$\lim_L [F_L^q(N_1), F_L^{-q'}(N_1)] = -\delta_{q,q'} \, sign(q) \tag{6.45}$$

$$\lim_L [F_L^q(N_2), F_L^{-q'}(N_2)] = \delta_{q,q'} \, sign(q) \tag{6.46}$$

Using the full central limit theorem Eq. (6.23), it puts us in a position to give an operator meaning to the density fluctuations in the thermodynamic limit for any value of the variable $q > 0$, namely as fluctuation operators. In particular we get the fluctuation $F^{-q}(N_1) = \lim_{L\to\infty} F_L^{-q}(N_1)$ and analogously for the other fluctuations $F^q(N_1), F^q(N_2), F^{-q}(N_2)$.

Using the commutation relations Eq. (6.45) and Eq. (6.46) we compute for any pair (λ, μ) of complex numbers

$$\widetilde{\omega}_0(e^{i\{\lambda F^{-q}(N_1) + \overline{\lambda}F^q(N_1) + \mu F^q(N_2) + \overline{\mu}F^{-q}(N_2)\}})$$
$$= \lim_L \omega_0(e^{i\{\lambda F_L^{-q}(N_1) + \overline{\lambda}F_L^q(N_1) + \mu F_L^q(N_2) + \overline{\mu}F_L^{-q}(N_2)\}}) \tag{6.47}$$
$$= e^{-\frac{1}{2}(|\lambda|^2 + |\mu|^2)}$$

where $\widetilde{\omega}_0$ is the macro-state determined by the central limit theorem. This macro-state is clearly a quasi-free state. It attaches a clear mathematical operator meaning to these fluctuations which are essentially the Fourier transforms of the density fluctuations. We created in all details the transition from the micro-system given by the triplet $\{\mathscr{A}_e, \omega_0, \alpha_t^0\}$ to the macro-system $\{\widetilde{\mathscr{A}}_e(q), \widetilde{\omega}_0, \widetilde{\alpha}_t^0\}$, where $\widetilde{\mathscr{A}}_e(q)$ stands for the algebra generated by the central limit q-density fluctuations Eq. (6.38),

Eq. (6.39). Also, $\tilde{\omega}_0$ is the state of these fluctuation algebra induced by the micro-state ω_0, and the transposed macro-dynamics $\tilde{\alpha}_t^0$ is induced by the free particle micro-dynamics α_t^0. We will now compute this dynamics $\tilde{\alpha}_t^0$ explicitly. Clearly the result Eq. (6.47) is obtained for each fixed but arbitrary value of the momentum q. On the basis of the computations Eq. (6.45), Eq. (6.46), it is reasonable to introduce the following notations: For all $q > 0$,

$$\alpha(q) = F^{-q}(N_1); \ \alpha(q)^* = F^q(N_1) \tag{6.48}$$
$$\beta(q) = F^q(N_2); \ \beta(q)^* = F^{-q}(N_2) \tag{6.49}$$

As a consequence of the commutators Eq. (6.45), Eq. (6.46), these macro-operators satisfy the boson canonical commutation relations

$$[\alpha(q), \alpha(q')^*] = [\beta(q), \beta(q')^*] = \delta_{q,q'}$$
$$[\alpha(q), \alpha(q')] = [\beta(q), \beta(q')] = 0$$

In these commutator relations we notice the presence of Kronecker δ functions and not δ-functions. The $\alpha(q)$, $\beta(q)$ and their adjoints are the creation and annihilation operators of two boson particles, acting on the GNS-representation space Eq. (7.1) of the macro-state $\tilde{\omega}_0$. This state is the boson Fock state as seen in Eq. (6.47). The transition from the physically relevant micro-fermion system to the macro-boson system of density fluctuations is completely realized.

We compute now the dynamics of the boson macro-system induced by the free micro-system dynamics. Let us first compute the following equation within the micro-system:

$$\delta_0(N_1(x)) \equiv \lim_L [H_L^0, N_1(x)] = i\frac{\partial}{\partial x} N_1(x)$$

and consider the fluctuation of this quantity in the limit $\hat{f}(k) \rightarrow 1/\sqrt{2\pi}$ for any $q > 0$. Using the boson creation and annihilation operator notation Eq. (6.48), we obtain the dynamical operator derivation $\tilde{\delta}_0$ of the macro-dynamics

$$\tilde{\delta}_0 \alpha^*(q) \equiv \lim_L F_L^q(\delta_0(N_1(x))) = q\alpha^*(q)$$

Analogously we obtain the relations

$$\tilde{\delta}_0 \alpha(q) = -q\alpha(q); \ \tilde{\delta}_0 \beta(q) = -q\beta(q); \ \tilde{\delta}_0 \alpha(q)^* = q\alpha(q)^*; \ \tilde{\delta}_0 \beta(q)^* = q\beta(q)^*$$

yielding the macroscopic free dynamics $\tilde{\alpha}_t^0 = e^{it\tilde{\delta}_0}$

$$\tilde{\alpha}_t^0(\alpha(q)) = e^{-iqt}\alpha(q) \text{ and } \tilde{\alpha}_t^0(\beta(q)) = e^{-iqt}\beta(q) \tag{6.50}$$

As a consequence of the time invariance of the free micro-state, the macro-state is also time-invariant and the derivation $\tilde{\delta}_0$ has a unique Hamiltonian (GNS-)representation given by $\tilde{\delta}_{0,q} = [\tilde{H}_{0,q}, .]$ where the Hamiltonian $\tilde{H}_{0,q}$ is computed straightforwardly to be

$$\widetilde{H}_{0,q} = q\{\alpha^*(q)\alpha(q) + \beta^*(q)\beta(q)\} \tag{6.51}$$

Hence we obtained a full representation of the unperturbed Luttinger micro-system lifted to the macro-level of the CCR-algebra of density fluctuations. The free micro-Hamiltonian is lifted to the fluctuation level Hamiltonian. In the literature [75] there seems to be a bit of a confusion about this point in the sense that these two Hamiltonians, namely the micro-Hamiltonian and the macro-Hamiltonian, are set equal to each other in an operator equality termed *the Kronig identity*. It is clear that in the thermodynamic limit this Kronig identity should not be understood as an identity of operators which are acting on the same Hilbert space. The micro-Hamiltonian acts on the fermion Hilbert space $\mathfrak{F}_{FaF}$, while the macro-Hamiltonian acts on a boson Fock Hilbert space, which is a completely different affair, a different world. Therefore the Kronig identity can hardly be considered as an operator identity.

In any case, as a first application of the theory of fluctuations a transition is created from the fermion ground state free Luttinger micro-subsystem $(\mathscr{A}_e, \alpha_t^0, \omega_0)$, to the free bosonic macro-system $(\widetilde{A}_{e,q}, \widetilde{\alpha}_t^0, \widetilde{\omega}_0)$ of q-density fluctuations.

Micro/macro-dynamics for the Interacting Luttinger Model

Now we switch on the interaction and look for the macro-dynamics of the density fluctuations generated by the full micro-dynamics given by the full Hamiltonian $H_L = H_L^0 + H_L^1$, where the interacting part of the Hamiltonian translates in Fourier language into

$$H_L^1 = \frac{2\lambda}{L} \sum_{p>0} v(p)\{\rho_{1,L}(-p)\rho_{2,L}(p) + \rho_{1,L}(p)\rho_{2,L}(-p)\} \tag{6.52}$$

Of course it is good to realize that the interaction terms depends only on the particle densities defined above. It is also important and interesting to notice that the interaction Hamiltonian is bilinear in these densities. The last property makes us consider the particular relation between the solvability of the system and the form of the interaction as being expressed in the creation and annihilation operators (see Chapter Eq. (5)). As we did for the free dynamics, we compute the time derivative for the full dynamics first within the micro-state ω_0, namely $\delta(N_1(x)) = \lim_L[H_L, N_1(x)]$; we then transport the result to the density fluctuations algebra using again the central limit technology. The total microscopic derivation $\delta = \delta_0 + \delta_1$ is straightforwardly computed and, applying the central limit theorems, induces another derivation $\widetilde{\delta}$ on the fluctuation CCR-algebra $\mathscr{A}_{e,q>0}$ which is, because of the bilinear character of the Hamiltonian in the densities, explicitly obtained to be

$$\widetilde{\delta}\alpha^*(q) = \lim_L F_L^q([H_L, N_1]) = q\{\alpha^*(q) + \frac{\lambda}{\pi}v(q)\beta(q)\}$$

$$\widetilde{\delta}\beta(q) = \lim_L F_L^q\{[H_L, N_2]\} = -q\{\beta(q) + \frac{\lambda}{\pi}v(q)\alpha(q)^*\}$$

One notices that the derivation $\widetilde{\delta}$ leaves the number of degrees of freedom of the macro-algebra $\widetilde{A}_{e,q}$ invariant. Again this derivation has a unique Hamiltonian representation in the state $\widetilde{\omega}_0$ given by $\widetilde{\delta} = [\widetilde{H}_q, .]$, where the fluctuations Luttinger Hamiltonian $\widetilde{H}_q$ is easily computed using the commutation relations of the $\alpha(q)$ and $\beta(q)$.

$$\widetilde{H}_q = q\{\alpha^*(q)\alpha(q)+\beta^*(q)\beta(q)\}+\frac{\lambda}{\pi}qv(q)\{\alpha^*(q)\beta^*(q)+\alpha(q)\beta(q)\} \quad (6.53)$$

This Hamiltonian is bilinear in the fluctuation macro-boson creation and annihilation operators and yields a solvable (see Chapter Eq. (5)) boson system of density-density interacting fluctuations. The corresponding total macro-dynamics is readily obtained in the boson Fock representation of the state $\widetilde{\omega}_0$ Eq. (6.47) given by $\widetilde{\alpha}_t(A) = e^{it\widetilde{H}_q}Ae^{-it\widetilde{H}_q}$ for any observable A, any element of the macro-algebra of the density observables $\widetilde{A}_{e,q}$.

Now we are interested in the derivation of the ground state of this interacting macro-system of density fluctuations. We must look for the solutions $\widetilde{\omega}$ of the ground state condition

$$\widetilde{\omega}(X^*[\widetilde{H}_q,X]) \geq 0$$

holding for all density fluctuations X, which are elements of $\widetilde{A}_{e,q}$. Note that the Hamiltonian $\widetilde{H}_q$ is the macro-image of the fully interacting micro-Hamiltonian acting on the representation space $\mathfrak{F}_{FaF}$.

This ground state problem is a solvable problem (see Eq. (5.1)) mainly because the Hamiltonian Eq. (6.53) is bilinear in the creation and annihilation operators $\alpha(q)$ and $\beta(q)$ Eq. (6.48). Solving this problem is merely a repetition of what one finds in the abundant literature about the Luttinger model as well as in Chapter Eq. (5) about solvable models. It consists essentially in diagonalizing the Hamiltonian $\widetilde{H}_q$, an operation which is performed by means of a boson Bogoliubov canonical transformation Eq. (7.3) acting on the macro-level Fock space. Acting on the latter space consider the unitary operator

$$U = \exp\{\varphi(q)(\alpha(q)\beta(q) - \alpha(q)^*\beta(q)^*)\}$$

where the real function φ is the solution of the equation: $\forall p$,

$$\tanh 2\varphi(p) = -\lambda\frac{v(p)}{\pi}$$

One checks that the conditions Eq. (6.35) are sufficient for the existence of such a solution. We obtain the diagonalized Hamiltonian in the form

$$U^*\widetilde{H}_qU = E(q)\{\widetilde{\alpha}^*(q)\widetilde{\alpha}(q)+\widetilde{\beta}^*(q)\widetilde{\beta}(q)\}$$

with $\widetilde{\alpha}(q) = U^*\alpha(q)U$ and, analogously, $\widetilde{\beta}(q) = U^*\beta(q)U$, the new quasi-particle creation and annihilation boson operators. Expressed in these operators the total Hamiltonian becomes equal to the sum of two free boson quasi-particles, each of these have the same one-particle energy function given by the function $E(p)$. There-fore the state $\widetilde{\omega}$ satisfying the ground state conditions is nothing more than the Fock state for these two boson quasi-particles. This state is related to the ground state $\widetilde{\omega}_0$ for the non-interacting free Luttinger model by the relation $\widetilde{\omega}(.) = \widetilde{\omega}_0(U.U^*)$. Furthermore we can easily compute the one-quasi-particle energy function

$$E(q) = |q|\sqrt{1 - (\frac{\lambda v(q)}{\pi})^2}$$

This function is called the one-particle spectral function of the collective excitations or of the relevant quasi-particles of the model. In the physics literature it is also sometimes called the dispersion relation of the model system.

We obtained the infinite-volume limit version of what is known in the literature as the exact solution of the Luttinger model. The limit behind this derivation is not one or other operator limit within the micro-model but something completely different, namely a central limit result which realizes a macroscopic system basically different from the original micro-system.

We presented this model also as an example of an explicit construction of the bosonization of a microscopic non-solvable fermion system. We can understand bet-ter what we mean by an exact ground state solution for the interacting Luttinger model. It deals with a rigorous construction of the macro-ground state for the macro-dynamics of the density fluctuations induced by the non-solvable micro-system. We do not speak about an exact ground state solution of the Luttinger micro-system.

Without going into the details, the non-solvability of the micro-model is due to the fact that we cannot find any closed finite set (the set of BBGKW-equations) of dynamical equations due to the presence of a non-trivial two-body fermion particle interaction, as explained in Chapter Eq. (5). An approximate procedure, well known in the physics literature, to bypass this kind of problems is to apply the so-called *random phase approximation* [133, 117]. Having applied the ideas and the results about quantum density fluctuation operators and all that around, however, we have been able to detect a solvable quasi-free macro-dynamics with a Hamiltonian bilinear in the density fluctuation creation and annihilation operators. In this sense we can consider this analysis as a proof that the random phase approximation becomes exact for the Luttinger model in the infinite-volume limit.

6.4.2 Micro/macro-dynamics and Conservation of Equilibrium

Suppose that we start with the micro-dynamical system given by the triplet $(\mathscr{A}, \omega, \alpha_t)$ with normal fluctuations, the situation discussed in section Eq. (6.2). The quantum central limit theorems map the dynamical micro-system onto the dynamical macro-system $(W(\mathscr{A}_{L,sa}, \sigma_\omega), \widetilde{\omega}, \widetilde{\alpha}_t)$ of its quantum fluctuations.

If the micro-state ω happens to be α_t-time invariant ($\omega \cdot \alpha_t = \omega$ for all $t \in \mathbb{R}$), then we have shown already that the macro-state $\widetilde{\omega}$ is also $\widetilde{\alpha}_t$-time invariant for the macro-dynamics, or expressed as a formula, $\widetilde{\omega} \cdot \widetilde{\alpha}_t = \widetilde{\omega}$ for all t.

We can ask a less trivial question: Suppose that the micro-state ω is an equilib-rium state for the micro-dynamics α_t—is the macro-state $\widetilde{\omega}$ also an equilibrium state for the macro-dynamics $\widetilde{\alpha}_t$ of the fluctuations? In [63] this question is answered positively although it is proved in a somewhat more technical sense than what we want to do here. In other words we do not reproduce all the technical details of the rigorous mathematical proof in [63]. Nevertheless we find it greatly instructive to get more details about this property by presenting a formal heuristic derivation of this result. To be explicit, we choose the EEB criterion Eq. (3.7) as the equilibrium criterion and assume that the micro-state ω satisfies this criterion for the local Hamiltonians $H_V(\mu)$. It follows from Eq. (3.8) that this state is micro-time invariant. As also mentioned above, the macro-state $\widetilde{\omega}$ is macro-time invariant. On the basis of the GNS-representation theorem (see Eq. (7.1)) for this macro-system, it has a Hamiltonian $\widetilde{H}(\mu)$ such that for any fluctuation $F(A)$),

$$\widetilde{\alpha}_t(F(A)) = F(\alpha_t A) = e^{it\widetilde{H}(\mu)} F(A) e^{-it\widetilde{H}(\mu)}$$

where the macro-Hamiltonian $\widetilde{H}(\mu)$ was defined before by

$$[\widetilde{H}(\mu), F(A)] = \lim_V F([H_V(\mu), A])$$

Using the EEB criterion for the micro-system, we obtain consecutively

$$\beta \widetilde{\omega}(F(A)^*[\widetilde{H}(\mu), F(A)]) = \lim_V \beta \, \omega(F_V(A)^*[H_V(\mu), F_V(A)])$$

$$\geq \lim_V \omega(F_V(A)^* F_V(A)) \ln \frac{\omega(F_V(A)^* F_V(A))}{\omega(F_V(A) F_V(A)^*)}$$

$$= \widetilde{\omega}(F(A)^* F(A)) \ln \frac{\widetilde{\omega}(F(A)^* F(A))}{\widetilde{\omega}(F(A) F(A)^*)}$$

expressing that the macro-state satisfies also the EEB criterion for the macro-dynamics at the same inverse temperature and chemical potential. This demonstrates the announced property.

This result proves the universal result that the notion of equilibrium is preserved under the operation of coarse graining induced by the central limit theorem. This statement constitutes a proof of one of the basic assumptions of the phenomenological theory of Onsager about small oscillations around equilibrium (see [62, 64] and references therein).

The preceding result contributes also to the discussion as to whether quantum systems should be described at the macroscopic level solely by classical observables. At least this opinion seems to have numerous believers. The fluctuation technology shows that the macroscopic fluctuation observables behave classically in all ways if they are constants of the motion. For observables which are not constants of the motion, this statement is false, In other words, a priori we can only expect that conserved quantities behave as classical observables. In principle, other macro-observables can behave as genuine quantum observables, keeping a genuine quantum dynamics. We have seen that quantum phenomena, like basic non-trivial canonical commutation relations behaviors, are observable at the macroscopic fluctuation level. We have seen in the Luttinger model analysis that the macro-dynamics is of the quantum mechanical type.

6.4.3 Micro/macro-dynamics and Linear Response Theory

Within the research area of equilibrium states, a standard well known procedure is to perturb the system by a gentle perturbation and to study the response of the system as a function of the perturbation. One learns that the response elucidates most (if not all) of the properties of the equilibrium state (see for instance [128]).

Technically speaking, we consider a perturbation of a given dynamics by adding a term to the originally stated Hamiltonian and we consider its perturbed dynamics. We expand these perturbed dynamics in terms of the perturbation and the unperturbed dynamics. Many will argue that when the perturbation is small we can limit the study of the response to the first order term in the perturbation series of the corresponding Dyson expansion [26]. This describes the essentials of the "linear response theory of Kubo" [94].

A long-standing debate has been going on about the range of validity of this linear response theory. The question remains about how to understand from a microscopic point of view the range of validity of the response theory being linear or intrinsically nonlinear. We should realize that the linear response actually observed in macroscopic systems seems to have a significantly greater range of validity, which goes far beyond the criticism being expressed about it.

We discuss the main result, which can be found in [65] where contours and setups are sketched, which are sufficient to guarantee the exactness of the linear response theory.

We assume that:

(i) the micro-dynamics α_t is again the limit of the local micro-dynamics $\alpha_t^V = e^{itH_V} \cdot e^{-itH_V}$, where again H_V contains only standard finite range interactions

(ii) the micro-state is a state of the form $\omega = \lim_V \omega_V$ with the property that it is time-translation invariant and ergodic, and that the local states ω_V are normal density matrix states

(iii) ω satisfies the clustering condition Eq. (6.20) for normal fluctuations and the equilibrium conditions Eq. (3.7) at a fixed inverse temperature β.

From the time invariance of the state ω we obtain a Hamiltonian H and a GNS-representation of the dynamics in the form $\alpha_t = e^{itH} \cdot e^{-itH}$. On the basis of the central limit theorems, we obtain the macro-dynamics $\widetilde{\alpha}_t$ of the fluctuation macro-algebra observables in the state $\widetilde{\omega}$. Its GNS-representation yields again a macro-Hamiltonian representation $\widetilde{H}$ of the form $\widetilde{\alpha}_t = e^{it\widetilde{H}} \cdot e^{-it\widetilde{H}}$.

Consider now any local micro-perturbation $P \in \mathscr{A}_{L,sa}$, and consider the following perturbed dynamics:

$$\alpha_{t,V}^P(.) = e^{it(H+F_V(P))} \cdot e^{-it(H+F_V(P))}$$

where $F_V(P)$ is the local fluctuation of the given micro-observable P. It is important to realize at this point that the perturbation is of the fluctuation type. In the physics literature one usually considers extensive perturbations. At any rate, by a direct computation we can show the following central limit theorem [65]: for all observables A and P in $\mathscr{A}_{L,sa}$; the perturbed macro-dynamics is of the following simple form:

$$\widetilde{\alpha}_t^P(.) = e^{it(\widetilde{H}+F(P))} \cdot e^{-it(\widetilde{H}+F(P))}$$

acting on the fluctuation macro-algebra in the sense

$$\widetilde{\alpha}_t^P F(A) = \lim_V F(\alpha_{t,V}^P(A))$$

and where $F(P)$ is the central limit of $F_V(P)$. This proves not only the existence, but reveals also the explicit form of the perturbed macro-dynamics at the level of the fluctuations. In particular we prove for each local observable A,

$$\lim_V \omega_V \left(\alpha_{t,V}^P F_V(A) \right) - \widetilde{\omega}(\widetilde{\alpha}_t^P F(A)) = 0$$

where $\widetilde{\omega}$ is the central limit macro-state. A formal check of this result is a straightforward exercise. This result is nothing more than the computation yielding the existence of the complete relaxation function of Kubo [94] lifted to the level of the fluctuations. It is of course important to always keep in mind that instead of the normal practice (namely dealing with extensive observables as perturbations) we consider here the fluctuations $F(P)$ as perturbations.

Furthermore, we have assumed that the micro-state ω is an (α_t, β)-equilibrium state. On the basis of the previous section, also the macro-state $\widetilde{\omega}$ on the fluctuations is an equilibrium state. From this we readily derive as an exact result Kubo's famous formula of his linear response theory, explicitly given by

$$\frac{d}{dt}\widetilde{\omega}(\widetilde{\alpha}_t^P F(A)) = i\widetilde{\omega}([F(P), \widetilde{\alpha}_t F(A)])$$

At this point, it is important to remark that the right hand side of this formula shows full linearity in the perturbation observable P. Kubo's formula appears as the transposed formula under the central limit of the microscopic response formula of the dynamics perturbed by a fluctuation observable. All this describes contours concerning the exactness of Kubo's linear response theory. Finally as $\widetilde{\omega}$ is an equilibrium state, the right hand side of the formula can be expressed in terms of the Duhamel two-point function [94], which is a common way of proceeding in linear response theory.

6.4.4 Micro/macro and SSB

Spontaneous symmetry breakdown(SSB) is in general one of the widespread and basic phenomena accompanying collective phenomena, such as phase transitions in statistical mechanics or interesting ground states in relativistic field theory. Spontaneous symmetry breakdown traces back to the Goldstone Theorem [68]. The existence or occurrence of SSB is one thing. For boson systems this is extensively studied in Section Eq. (4.3.2). The study of its properties (i.e., the study of specific or typical properties of symmetry breaking states) is another thing. In the latter case, there are several different situations to be considered that yield different characteristic properties. In particular, for instance in the case of short range interactions, it is typical that SSB yields a dynamics that maintains the symmetry and shows a gapless energy spectrum. This is the situation of the Goldstone theorem [68]. On the other hand for long-range interactions, SSB breaks the symmetry of the dynamics and for lattice systems (see below) yields a spectral gap This gap is accompanied by new system oscillations with a spectral frequency taking a finite value at momentum $k = 0$ [132, 89, 36]. However, what seems to be a general feature of spontaneous symmetry breaking in all cases is that the physics literature predicts the appearance of a particular particle, called the Goldstone-Nambu boson, which appears necessarily as a consequence of SSB. If SSB disappears also this boson seems to disappear. At this point, we maintain that the theory of fluctuation operators allows for the explicit construction of the field canonical coordinates of this Goldstone particle. The most general results about this feature can be found in [124] and its applications to boson systems in [125]. We sketch the essentials for quantum spin systems in the two cases. First, we treat micro-systems with long-range interactions, (in fact, for quantum spin mean field systems) and afterwards we consider micro-systems with short range interactions.

Long Range (Mean Field) Interactions

As an example of the long range interaction situation, we discuss the Anderson version of the strong coupling BCS (Bardeen-Cooper-Schriefer) model in one dimension $(d = 1)$. This model is considered to be the prototype Hamiltonian model explaining the condensation of the Cooper pairs of electrons. These electron pairs systems behave as a quasi-local subsystem of the complete fermion system. Therefore it allows for the application of the quantum fluctuation theory. This model is at the basis of the microscopic interpretations of the phenomenon of superconductivity. Results similar to the ones given for this model are found in [28, 29, 30] for other models.

This microscopic model of a BCS-system is again given by the spin algebra of observables $\mathscr{A}_L = \otimes_i (M_2)_i$ where M_2 is the algebra of 2×2 complex matrices. The local Hamiltonian H_N of the model for finite N is given by

$$H_N = \varepsilon \sum_{i=-N}^{N} \sigma_i^z - \frac{1}{2N+1} \sum_{i,j=-N}^{N} \sigma_i^+ \sigma_j^- \; ; \; 0 < \varepsilon < \frac{1}{2}$$

where $\sigma^z, \sigma^\pm$ are the usual 2×2-Pauli matrices. The operators σ^z stand for the number operators of the Cooper pairs of particles near the Fermi sea. The operators $\sigma^\pm$ are the creation and annihilation operators of these pairs. Because of the mean field character of the interaction in the thermodynamic limit $(N \to \infty)$ the equilibrium conditions Eq. (3.7) have the following ergodic product state solutions (see [50]): $\omega_\lambda = \otimes_i tr \rho_{\lambda,i}$, where the 2×2-density matrices $\rho_{\lambda,i} = \rho_\lambda$ at each lattice site $\{i\}$ with ρ_λ a solution of the self-consistency equation for this density matrix

$$\rho_\lambda = \frac{e^{-\beta h_\lambda}}{tr \, e^{-\beta h_\lambda}}$$

where

$$h_\lambda = \varepsilon \sigma^z - \lambda \sigma^+ - \bar{\lambda} \sigma^-$$

This equation is equivalent with the equation for the parameter λ, namely $\lambda = tr \rho_\lambda \sigma^- = \omega_\lambda(\sigma^-)$. The parameter λ is often called the *order parameter* of the model. Obviously, σ^- is the corresponding order parameter operator.

We should note that $\lambda = tr \rho_\lambda \sigma^-$ is a nonlinear equation for λ whose solutions determine the operator h_λ and therefore the density matrix ρ_λ and finally the equilibrium state ω_λ. The equation always possesses the solution $\lambda = 0$, describing the so-called normal phase of non-interacting Cooper pairs. For $\beta > \beta_c$, where β_c is the solution of the equation $th\beta_c \varepsilon = 2\varepsilon$, we obtain a solution $\lambda \neq 0$, describing the superconducting phase. It is clear that β_c is the inverse critical temperature. If λ is a non-zero solution, then also $\lambda e^{i\varphi}$ is a non-zero solution for all real φ. Hence if there exists one solution $\lambda \neq 0$, then we immediately obtain infinitely many solutions.

Clearly the Hamiltonian H_N is invariant under the continuous gauge canonical transformations group $\mathscr{G} = \{\gamma_\varphi | \varphi \in [0, 2\pi]\}$ of the micro-algebra defined by the action

$$\gamma_\varphi(\sigma_i^+) = e^{-i\varphi} \sigma_i^+$$

The group $\mathscr{G}$ is a symmetry group of the Hamiltonian system H_N. On the other hand, for any $\lambda \neq 0$ we find for $\varphi \neq 0$ that

$$\omega_\lambda(\gamma_\varphi(\sigma_i^+)) = e^{-i\varphi} \omega_\lambda(\sigma_i^+) \neq \omega_\lambda(\sigma_i^+).$$

This means that the equilibrium conditions Eq. (3.7) produce an infinite number of solutions, all breaking the symmetry. The gauge group $\mathscr{G}$ is spontaneously broken. We note that the gauge transformations are locally implemented by the generators of the symmetry given by $Q_N = \sum_{j=-N}^{N} \sigma_i^z$ for large enough N. In other words, $\gamma_\varphi(\sigma_i^+) = e^{-i\varphi Q_N} \sigma_i^+ e^{i\varphi Q_N}$ or Q_N is the generator of the symmetry group and σ^z is the symmetry generator density. As all the states ω_λ are product states, they satisfy the scaling condition Eq. (6.20) and all fluctuations are normal. Omitting the locality index, we consider the local operators

$$Q = \frac{|\lambda|^2}{\mu^2} \sigma^z + \frac{\varepsilon}{\mu^2} (\lambda \sigma^+ + \bar{\lambda} \sigma^-)$$

$$P = \frac{i}{\mu} (\lambda \sigma^+ - \bar{\lambda} \sigma^-)$$

where $\mu = (\varepsilon^2 + |\lambda|^2)^{1/2}$. Note that P is essentially the normalized self-adjoint part of the order-parameter operator, the operator indicating the symmetry breaking of the states, expressed by

$$\frac{d}{d\varphi} \omega_\lambda(\gamma_\varphi(P)) \neq 0 \text{ and }, \omega_\lambda(P) = 0$$

On the other hand Q is given mainly by the generator of the symmetry σ^z plus a self-adjoint part coming from the normalization to zero, which also means that $\omega_\lambda(Q) = 0$.

It is a straightforward computation to check (see also [124, 125] for details), that the fluctuations $F(Q)$ and $F(P)$ Eq. (6.23) form a canonical pair satisfying the boson commutation relation

$$[F(Q), F(P)] = i\frac{4|\lambda|^2}{\mu} \tag{6.54}$$

a relation which holds as an equation acting on the GNS-representation space for the fluctuations determined by the state $\widetilde{\omega}_\lambda$ obtained from the central limit theorem applied to the original micro-system state ω_λ.

These two fluctuation operators behave under the macro-dynamical time evolution $\widetilde{\alpha}_t$ Eq. (6.22) as harmonic oscillator (q,p)-coordinates oscillating with a frequency equal to 2μ. This frequency is often called the *plasmon frequency* of the system. We recognize the factor 2, attributed to the well-known frequency doubling in the BCS-model.

One computes as well the variances of the two operators $F(P)$ and $F(Q)$ and obtain the equalities

$$\widetilde{\omega}_\lambda(F(Q)^2) = \frac{|\lambda|^2}{\mu^2} = \widetilde{\omega}_\lambda(F(P)^2)$$

This expression tells us that these fluctuation coordinates vanish, hence disappear, if the order parameter $\lambda = 0$ vanishes. The operator coordinates $F(Q)$ and $F(P)$ are non-trivial canonical coordinates of a particle appearing if and only if there is spontaneous symmetry breakdown. In fact they are the so-called *normal coordinates* of the Goldstone-Nambu boson, which show up if and only if SSB occurs.

Short-range Interactions

Results comparable with those in the case of long range interactions can be derived for systems with short range interactions. The main difference between the two cases is that, in the short-range interactions case, SSB yields equilibrium states with bad mixing properties [116, 48]. The correlation functions slowly decrease at infinity; in general, they are not integrable. Therefore we arrive at a situation comparable to the one described in the section that discusses abnormal fluctuations. However, in this case one also encounters the phenomenon that SSB manifests the appearance of a Goldstone particle. We can also explicitly construct its canonical coordinates. The details of this construction for more or less arbitrary systems are found in [125], whereas the results typical for bosons are found in [124].

Here we provide a formal picture of this construction for quantum spin microsystems; we are not discussing all mathematical details of the construction. We consider again the prototype spin micro-system $(\mathscr{A}_L, \omega, \alpha_t)$. Let $\{\gamma_s | s\}$ be a strongly continuous, one-parameter symmetry group of the micro-dynamic system α_t, which is locally generated by the generator, (sometimes also called the *charge operator*, especially in high-energy physics), $Q_V = \sum_{x \in V} q_x$, with q_x called the local density of

Q_V and $\gamma_s(X) = \lim_V e^{isQ_V} X e^{-iQ_V}$ for each observable X. Spontaneous symmetry breaking amounts to finding an equilibrium or ground state ω that breaks the symmetry $\{\gamma_s\}$ (see Eq. (4.3.2)). This means that there exists at least one local observable $A \in \mathcal{A}_{L,sa}$ such that $\omega(\gamma_s(A)) \neq \omega(A)$ for $s \neq 0$, while nevertheless $\alpha_t \gamma_s = \gamma_s \alpha_t$. The last remark expresses the fact that the symmetry of the dynamics is not broken, contrary to the long-range interaction case. As before, the observable A is sometimes called the *order parameter operator* and its expectation value $\omega(A)$ the order parameter.

We can easily argue that this property is equivalent to the existence of a complex number $c \neq 0$ (we can also show that this constant is time independent) such that

$$\frac{d}{ds}\omega(\gamma_s(A))\Big|_{s=0} = \lim_V \omega([Q_V, A]) = c \neq 0$$

By carrying out a number of trivial steps we can turn this form of the SSB-definition equation into a commutator relation for fluctuations. Using the space translation invariance of the state, we obtain the commutator expression

$$\lim_V \frac{1}{V} \omega\left(\left[\sum_{x \in V}(q_x - \omega(q)),\ \sum_{y \in V}(\tau_y A - \omega(A))\right]\right) = c$$

Now we implement the other consequence of the spontaneous symmetry breaking [116, 48], namely the fact that SSB for short-range interactions implies bad clustering properties, as discussed before also sometimes called *off-diagonal long range order*, for the order parameter operator A. We formalize this property in the following particular form: We assume that the lack of clustering can be expressed by the existence of a strictly positive index δ, a real number such that the limit

$$0 < \lim_V \omega\left(\frac{1}{V^{1+2\delta}}\left(\sum_{x \in V}(\tau_x A - \omega(A))\right)^2\right) < \infty$$

exists, is nontrivial, and is finite. This is a simple manner of expressing that the abnormal fluctuation

$$F^\delta(A) \equiv \lim_V \frac{1}{V^{1/2+\delta}} \sum_{x \in V}(\tau_x A - \omega(A))$$

does exist, as seen in Eq. (6.3). We write our commutator expressing, namely the SSB-property, as follows

$$\lim_V \omega\left(\left[\frac{1}{V^{1/2-\delta}}\sum_{x \in V}(q_x - \omega(q)),\ \frac{1}{V^{1/2+\delta}}\sum_{y \in V}(\tau_y A - \omega(A))\right]\right) = c$$

yielding in the limit the fluctuation commutator expression

$$\widetilde{\omega}\left(\left[F^{-\delta}(q),\ F^\delta(A)\right]\right) = c.$$

If the state ω is the equilibrium micro-state and hence its induced state $\tilde{\omega}$ is an equilibrium state of the fluctuations, we get the operator equation of two fluctuations

$$[F^{-\delta}(q) , F^{\delta}(A)] = c\,1.$$

This is because each equilibrium state is a faithful state (see [26]). In other words, we obtain the canonical pair $(F^{-\delta}(q) , F^{\delta}(A))$ of the quantum normal coordinates of a particle. It is not a bare particle, its coordinates are the collective quantities $(F^{-\delta}(q), F^{\delta}(A))$ of fluctuations. We found the new particle, the Goldstone-Nambu quasi-particle behind the phenomenon of SBB.

We should note that the property of long-range correlations for the order-parameter operator is the reason for the strict positivity of the parameter δ, and that the latter one is exactly compensated by a squeezing phenomenon of exactly the same strength described by the strictly negative index $-\delta$ for the fluctuation operator of the local generator q_x of the broken symmetry. This result must be considered as a typical phenomenon related to spontaneous symmetry breaking versus non-spontaneous symmetry breaking or versus externally forced breaking of the symmetry. In the case of SSB, this squeezing phenomenon is sometimes explained by the phrase that the symmetry is not completely broken, but only partially broken in the sense that the fluctuations of the generators of the symmetries, which are spontaneously broken, are softer than normal or are super-normal fluctuations. The property is mathematically expressed by the formula $-1/2 < -\delta < 0$. This interesting phenomenon was described long ago in a completely different context for some solvable solid state models in [4, 157]. More detailed rigorous information about all this is found in references [124, 125].

SSB and Fluctuations in Continuous Boson Systems

Of course the fluctuation theory is applicable as well if the micro-system is already a boson system. In the analysis above we describe the crucial role played by the SSB phenomenon. In Eq. (4.3.2) we proved that SSB is present in boson systems if and only if there is Bose-Einstein condensation. In the following discussion we consider only closed homogeneous boson systems with condensation. In that case we have an equilibrium state with a strictly positive zero-mode condensate density $\rho_0 > 0$ and the gauge symmetry is spontaneously broken. The total particle number operator $Q_V = N_V = \int_V dx\, a^*(x)a(x)$ is the generator of the continuous gauge group. The generator is the integral of the local density operator $n(x) = a^*(x)a(x)$. From Eq. (4.3.2) we know that the limit Gibbs states, (i.e., any homogeneous gauge invariant equilibrium state), which we denoted by ω_β, can be written as an integral over ergodic symmetry breaking equilibrium states $\{\omega_\beta^\varphi \,|\, \varphi \in [0, 2\pi]\}$ in the following form: For any observable X,

$$\omega_\beta(X) = \frac{1}{2\pi} \int_0^{2\pi} d\varphi\; \omega_\beta^\varphi(X)$$

We also derived that each of the ergodic states ω_β^φ satisfies the property

$$\lim_V \omega_\beta^\varphi (\frac{a_0}{\sqrt{V}}) = \lim_V \omega_\beta^\varphi (\frac{1}{V} \int_V dx\, a(x)) = \sqrt{\rho_0}\, e^{-i\varphi}$$

Without loss of generality, we consider for simplicity the phase $\varphi = 0$. The local order-parameter operator for boson condensation is given by the creation or annihilation operators $a^*(x)$ or $a(x)$. In practice it is always wise to work with a self-adjoint linear combination of these operators; therefore we use for the local order parameter operator A the linear combinations $A = \frac{i}{2}(a^*(x) - a(x))$ or $A = \frac{1}{2}(a^*(x) + a(x))$. From our analysis above it follows that the local density operator has a squeezed number fluctuation operator $F^{-\delta}(n(x))$, while the local order-parameter operator A has an abnormal fluctuations $F\delta(A)$. The degrees (the values of the corresponding δ's) of squeezing and abnormality depend on the model and on the chosen boundary conditions. For periodic boundary conditions we find detailed results for the mean field boson gas and for the Bogoliubov model in [124]. In all cases we obtain the following commutation relation for the canonical pair of the density and the order-parameter fluctuation operators:

$$[F_{-\delta}(n), F_\delta(A)] = i\sqrt{2\rho_0} \tag{6.55}$$

We finish our discussion on boson systems by making some remarks about what is called the problem of the *quantum phase operator*. The belief in the existence of such an operator occurred in a natural way in the field of coherent radiation, one of the most exciting inventions of the 20th century. It was at the origin of the masers and the lasers before they became regular features in daily life. (We think about our micro-wave ovens.)

Coherent radiation not only produces high intensity photon radiation but it also radiates all photons in a coherent phase. Characteristic for such coherent states Eq. (2.22) is that they are typical for quantum phenomena such as that we might find within Bose-Einstein condensation, superfluidity, and superconductivity. It is important to realize that all these phenomena are by now understood as being quantum collective phenomena, in other words quantum macroscopic effects.

Since the early days of quantum mechanics, physicists tried to look for the genuine quantum variables for these coherent phase phenomena. The quantum mechanics literature informs us about the long history describing the attempts, in particular those attempts of formulating the quantum phase operator as the canonical dual operator of the number operator N acting on the Fock space level. Unfortunately, none of these attempts were successful. Indeed all of them suffered from the following simple mathematical contradiction: Let Θ be any self-adjoint operator, which is the canonical dual of the boson number operator N both acting on the Fock space. Hence, we suppose that the operator Θ satisfies the commutator relation $[\Theta, N] = i1$, with all the operators acting on the Fock space $\mathfrak{F}$. We now take the Fock state expectation value of this commutation relation, that is, we sandwich this relation left and right between the Fock vacuum vector Ω_F. Because of the vacuum property of this Fock vector $N\Omega_F = 0$ we immediately obtain $0 = (\Omega_F, [\Theta, N]\Omega_F) = i$, which yields of course a nonsensical contradiction. This shows that the operator Θ as the dual of the number operator cannot exist as an operator satisfying the commutation relation when acting on the Fock space. Of course this simple argument leaves open many technical gates in order to try to escape this deadlock. So far, however, all candidate escape routes, or all models trying to give a decent answer to the formulation of a quantum phase operator within a Fock space formulation, have their own serious deficiencies. For an extensive literature and discussion about these models and topics we refer to, for example, [40] and references therein. It is fair to say that Fock space considerations did not yet offer any acceptable formulation for carrying the presence of the quantum phase operator. Faced with this fact and the knowledge that the boson Fock space is able to describe only phenomena of zero density of particles, it is reasonable to conclude that the phenomena of Bose-Einstein condensation, superconductivity or superfluidity also cannot be fully understood solely on the basis of Fock space considerations, exactly because they are all collective or macroscopic phenomena. Moreover as pointed out before, it is typical that these effects are always accompanied by spontaneous symmetry breaking. We note however that we meet all the prerequisites for having a quantum phase operator in the result obtained within Eq. (6.55). There, we have obtained a canonical dual pair of macroscopic variables consisting of $F_{-\delta}(n)$, namely the number operator fluctuation and as its dual operator, the order parameter fluctuation operator $F_\delta(A)$. Furthermore, both operators are macro-observables. Therefore the operator $F_\delta(A)$ plays perfectly the role of the quantum phase operator of the problem. We may conclude that this application of the general fluctuation theory yields a solution for a long standing problem about the existence of the quantum phase operator and its place in a quantum theory.

Ultimately it is also interesting to point out another property of the number operator fluctuation on the level of the fluctuations algebra. We check immediately that the operator $F_{-\delta}(n)$ implements the gauge transformations group $\{\widetilde{\gamma}_\varphi / \varphi \in [0, 2\pi]\}$ on the macroscopic level in the following sense:

$$\widetilde{\gamma}_\varphi(F(A)) = e^{i\varphi F_{-\delta}(n)} F(A) e^{-i\varphi F_{-\delta}(n)}$$

and

$$\widetilde{\omega}_\beta^{\varphi_1 + \varphi_2}(\,.\,) = \widetilde{\omega}_\beta^{\varphi_1} \circ \widetilde{\gamma}_{\varphi_2}(\,.\,)$$

acting on any reasonable fluctuation operator. In other words the phase operator is the generator of the spontaneously broken gauge group mapping each ergodic gauge breaking state into each other one. That is exactly what we ask a phase operator to do.

These phase operator remarks are not special only for boson systems. They hold as well for the BCS-model, a long range interacting model. All these properties follow from the canonical commutation relation Eq. (6.55). Such an analogous equation is written down for continuous fermions systems, for quantum spin systems in general, and so on. In all these situations, if coherence comes around the corner, such a quantum phase operator might show up on the level of the fluctuations yielding a solution for this long standing question.

7

Appendix

7.1 Dynamical Systems and GNS Construction

There exists a formulation of a dynamical physical system that is particularly suitable for the handling of systems with infinitely many degrees of freedom, which is the case after having taken the thermodynamic limit. It is particularly handy for the study of the collective phenomena. This formulation is sometimes called the algebraic approach to statistical mechanics and field theory. Extensive mathematical and technical information about this approach can be found in [26]. For a book with a view focused on the applications in many body systems we refer, as one example, to [152].

Trying to put in the picture only the main ideas of this theory, which are valuable in the applications, we present a proper version of the definition of a dynamical physical system. It is generally valid, it can be a quantum or a classical system. In this Appendix, we focus on the generalized Hilbert space representation theory of dynamical systems. It is called the GNS (Gelfand-Naimark-Segal)-construction.

Dynamical Physical System

After a moment of cogitation we should convince ourselves that a *dynamical physical system* is essentially given by a triplet of the type $(\mathfrak{A}, \mathscr{E}, (\alpha_t)_t)$, where $\mathfrak{A}$ stands for the set of relevant observables of the system under consideration, $\mathscr{E}$ for the set of states of the system, and $(\alpha_t)_t$ for the dynamics of the system. The real t is the time parameter.

First we discuss the identification of the three physical objects in the triplet and we find their mathematical characterizations.

We start considering the set of observables $\mathfrak{A}$. Keeping in mind, for instance ordinary quantum mechanics, an observable is a linear operator acting on a Hilbert space. As examples, we can consider the position operator x and the momentum operator p. Other observable quantities like angular momentum, energy, and so on, are again linear operators constructed out of linear combinations of products of the

A.F. Verbeure, *Many-Body Boson Systems*, Theoretical and Mathematical Physics,   165
DOI 10.1007/978-0-85729-109-7_7, © Springer-Verlag London Limited 2011

position x and the momentum p. We conclude then that it is reasonable to assume that the general notion of a set, denoted $\mathfrak{A}$, of observable quantities should be an *algebra*, that is, a linear space endowed with a product rule. A physically measurable observable should have a real spectrum and therefore there should be on the algebra a generalized notion of self-adjointness which is linked to the notion of conjugation. At this general level we speak in mathematics about an *involution*. An involution on $\mathfrak{A}$ is a map, indicated by .*, which maps any observable $X \in \mathfrak{A}$ into another one denoted by $X^* \in \mathfrak{A}$, such that $\forall X, Y \in \mathfrak{A}$, $\forall \lambda \in \mathbb{C}$ holds $(X^*)^* = X$, and $(\lambda X)^* = \bar{\lambda} X^*$, $(XY)^* = Y^* X^*$ and finally $(X + Y)^* = X^* + Y^*$. We assume that any acceptable algebra of observables $\mathfrak{A}$ has such an involution. We will find concrete examples of involutions of the boson algebra and of the spin system algebra throughout many chapters in this book.

For our purposes we assume also that the algebra contains a unit element, that is, an element $\mathbf{1}$ with the property that $\mathbf{1}A = A\mathbf{1} = A$ for all $A \in \mathfrak{A}$.

Many times to prove actually a number of items, in particular continuity properties, it is convenient that the algebra $\mathfrak{A}$ is a normed algebra with involution, that is, that there exists also a *norm* on $\mathfrak{A}$, that is, a map $||.||$ from $\mathfrak{A}$ into the non-negative real numbers such that $\forall X, Y$: $||X|| > 0$ if $X \neq 0$, $||\lambda X|| = |\lambda| \, ||X||$, $||X + Y|| \leq ||X|| + ||Y||$, $||XY|| \leq ||X|| \, ||Y||$ and $||X^*|| = ||X||$.

If a normed algebra with involution is complete with respect to its norm topology, we speak of a *Banach algebra with involution*.

The so-called algebraic approach to statistical mechanics and field theory [26] is based on the fact that the algebra of observables $\mathfrak{A}$ is a Banach algebra with involution. It is usually assumed that it has moreover the C^*-property. This property means that $||X^*X|| = ||X||^2$ for each observable X. Again the latter property makes a number of mathematical properties and proofs possible, but on the other hand it also limits the applicability of the notion of algebra of observables to realistic physical systems. We are not assuming a priory that the algebra of observables has the C^*-property.

We should realize that our definition of algebra of observables is also valid for classical systems. Of course in the latter case, the algebraic product is commutative.

Next we discuss a number of general properties about the notion of *state* and the notion of the *set of states $\mathscr{E}$*.

Any positive linear normalized functional on the algebra of observables $\mathfrak{A}$ is called a state. In particular a state ω assigns to each observable $X \in \mathfrak{A}$ a complex number $\omega(X) \in \mathbb{C}$ satisfying the following three properties:

(i) the linearity, for $\lambda, \mu \in \mathbb{C}$, and $X, Y \in \mathfrak{A}$ holds, $\omega(\lambda X + \mu Y) = \lambda \, \omega(X) + \mu \, \omega(Y)$

(ii) the positivity, $\omega(X^*X) \geq 0$ for all $X \in \mathfrak{A}$

(iii) the normalization, $\omega(\mathbf{1}) = 1$.

We should realize immediately that a state has all essential properties of an expectation-valued map. Therefore it is a natural generalization to the non-commutative systems of the notion of expectation in probability or in classical mechanical systems.

It is nice to remark that for each real $\lambda \in [0,1]$ and for each pair of states $\omega_1, \omega_2 \in \mathscr{E}$ holds that the convex combination $\omega = \lambda \omega_1 + (1-\lambda) \omega_2$ is again a state of the system. We therefore conclude that the set of states $\mathscr{E}$ is a *convex set* in this sense.

Starting from an algebra $\mathfrak{A}$ with an involution and a state ω on the algebra, then the following general properties hold:

1. For all $X, Y \in \mathfrak{A}: \overline{\omega(X^*Y)} = \omega(Y^*X)$
2. $|\omega(X^*Y)|^2 \leq \omega(X^*X)\omega(Y^*Y)$
3. $\omega(X^*) = \overline{\omega(X)}$
4. $|\omega(X)|^2 \leq \omega(X^*X)$

As a matter of getting acquainted to the notion of a state, it is good to prove explicitly these general properties. The proof of these properties is mainly based on the simple remark that the state ω defines something which is almost an inner product, sometimes called a pre-scalar product $(.,.)$, defined on the vector space underlying the algebra $\mathfrak{A}$. It is defined by

$$\forall X, Y \in \mathfrak{A}, \ (X,Y) \equiv \omega(X^*Y) \tag{7.1}$$

The definition of this pre-scalar product will also be the cornerstone for the GNS-construction which we discuss later. Properties 1 and 3 follow immediately from the positivity of the state ω. Properties 2 and 4 are direct consequences of the Schwartz inequality $|(X,Y)|^2 \leq (X,X)(Y,Y)$.

If the algebra of observables is moreover a Banach algebra with involution, we also obtain the norm continuity of every state. This means that for each state ω and for each observable $X \in \mathfrak{A}$,

$$|\omega(X)| \leq ||X|| \tag{7.2}$$

Needless to stress here the importance of this property. It means that it is sufficient to specify the state on a norm dense subset of the set of observables in order to know the state everywhere on the closure of the set.

Now we proceed to the proof of this continuity property, which goes as follows: Take any observable X such that $X^*X < 1$. Then use the fact that the series expansion of the complex function $\sqrt{1-z}$, $z \in \mathbb{C}$ around the point $z = 0$ is absolutely convergent for $|z| < 1$. Therefore we obtain the norm convergent series expansion for $\eta = \eta^* \in \mathfrak{A}$, where $\eta^2 = \eta^*\eta = 1 - X^*X$, given by

$$\eta = \sqrt{1-X^*X} = 1 - \frac{X^*X}{2} + (\frac{1}{2})^2 \frac{(X^*X)^2}{2!} - (\frac{1}{2})^3 \frac{3}{2} \frac{(X^*X)^3}{3!} + \cdots$$

The positivity of the state ω yields $0 \leq \omega(\eta^*\eta) = \omega(\eta^2) = 1 - \omega(X^*X)$.

Take now an arbitrary observable X and let $\widetilde{X} = \frac{X}{||X||+\varepsilon}$ for an arbitrary real number $\varepsilon > 0$. Then $||\widetilde{X}|| \leq 1$ or $\widetilde{X}^*\widetilde{X} < 1$, hence $\omega(\widetilde{X}^*\widetilde{X}) < 1$ or equivalently $\omega(X^*X) \leq (||X||+\varepsilon)^2$. Finally let ε tend to zero. We obtain the above continuity property of the state. This completes the proof of all the mentioned basic properties of an arbitrary state.

Finally we discuss the *dynamics*, the third item in the definition of a dynamical physical system. The notion of a reversible dynamics is mathematically translated to that of a *-automorphism of the algebra of observables. In boson physics such an automorphism is generally called a *canonical transformation*. See also Eq. (7.3). In words it is essentially a transformation of the creation and annihilation operators into new ones leaving the canonical commutation relations invariant. For our purpose we continue using the nomenclature of canonical transformation instead of that of *-automorphism. Mathematically, such a map denoted by τ, is a one-to-one map of $\mathfrak{A}$ onto $\mathfrak{A}$ such that, for all $X, Y \in \mathfrak{A}$ and $\lambda \in \mathbb{C}$, respectively $\tau(XY) = \tau(X)\tau(Y)$, $\tau(X + Y) = \tau(X) + \tau(Y)$, $\tau(\lambda X) = \lambda \tau(X)$, $\tau(X)^* = \tau(X^*)$. If $\mathfrak{A}$ is an algebra with the C^*-property, we need to add the property $\|\tau(X)\| = \|X\|$.

A *reversible process or dynamics* is given by a one-parameter group $(\alpha_t)_{t \in \mathbb{R}}$ of canonical transformations α_t. This means that $\forall t, s \in \mathbb{R}$ holds $\alpha_t \alpha_s = \alpha_{t+s}$, $(\alpha_t)^{-1} = \alpha_{-t}$, $\alpha_0 = id$, $\alpha_t(\mathbf{1}) = \mathbf{1}$. Naturally, the parameter t stands for the time variable.

On the other hand an *irreversible process or dynamics* is described by a one-parameter semigroup $(\alpha_t)_{t \in \mathbb{R}^+}$ of linear, positive, unity conserving maps α_t of the algebra of observables $\mathfrak{A}$. That is, $\forall t, s \in \mathbb{R}^+$ holds $\alpha_0 = id$, $\alpha_t \alpha_s = \alpha_{t+s}$, α_t is linear, $\alpha_t(X^*X) \geq 0$, and $\alpha_t(\mathbf{1}) = \mathbf{1}$.

We need to understand that the presentation of the dynamics is a generalization to general dynamical systems of what is known in quantum mechanics as the Heisenberg picture of the time evolution. The so-called Schrödinger picture is easily obtained from this Heisenberg picture after transportation of this dynamics to the state space $\mathcal{E}$. That goes as follows: Let ω be any state. Define the map $\widetilde{\alpha}_t$ on the state space, that is, on any element ω by $\widetilde{\alpha}_t(\omega) = \omega \circ \alpha_t$. This $\widetilde{\alpha}_t$, mapping $\mathcal{E}$ into itself, is the corresponding Schrödinger evolution. Clearly for this generalized setup of dynamical system, going from the Heisenberg picture to the Schrödinger picture is straightforward. The inverse move is a somewhat more delicate operation. We do not enter here any further details about this problem.

A state ω is called *stationary or time invariant* if the equality $\omega \circ \alpha_t = \omega$ holds for all values of the time parameter t.

In this section we continue our discussion with more details about the reversible dynamics and leave the irreversible ones for the next Section Eq. (7.2).

GNS-Construction

indexGNS-construction A *representation* of the algebra $\mathfrak{A}$ on a Hilbert space $\mathcal{H}$ is a morphism π of the algebra into the possibly unbounded linear operators $\mathcal{L}(\mathcal{H})$ defined on a dense domain of $\mathcal{H}$. Hence $\pi : X \to \pi(X)$ such that π is linear, commutes with the involution $\pi(X^*) = \pi(X)^*$, and conserves the product rules $\pi(XY) = \pi(X)\pi(Y)$.

Conversely, if we have such a representation π of the algebra on a Hilbert space $\mathcal{H}$, and we take any normalized vector ξ of $\mathcal{H}$, then the expectation-valued map ω_ξ defined by

$$\omega_\xi(X) = (\xi, \pi(X)\xi)$$

is a state on the algebra determined by the given representation and the given specific vector ξ. Hence any representation π of the algebra defines (many) states (expectation-valued maps) on the algebra.

Now we are interested in the inverse operation. We show how to get a representation of an algebra of observables given an arbitrary state on the algebra. The theorem which illustrates this possibility by explicit construction of this representation is called the Gelfand-Naimark-Segal, or simply the GNS-construction (see [26] for much more mathematical details).

This theorem is usually formulated for states on Banach algebras. For Bose systems this means that we must work with the Weyl-algebra of observables, which is generated by the unitary Weyl-operators $W(f), f \in \mathscr{S}$, with $\mathscr{S}$ the smooth and rapidly decreasing complex-valued functions on $\mathbb{R}^d$. The Weyl operators are considered as acting on the Fock space. The involution coincides of course with taking the adjoint with respect to the scalar product of the Fock space and the norm coincides with the usual operator norm. We are ready to formulate the main theorem.

Theorem 7.1. *Let $\mathfrak{A}$ be a Banach algebra with involution and identity and let ω be any state on it. Then there exists a Hilbert space $\mathscr{H}_\omega$, a representation π_ω of $\mathfrak{A}$ into the bounded operators $\mathscr{B}(\mathscr{H}_\omega)$ on $\mathscr{H}_\omega$. Also, there exists a (cyclic) vector Ω_ω such that:*

(a) $\mathscr{H}_\omega$ is generated by the set of vectors $\{\pi(X)\Omega_\omega | \forall X \in \mathfrak{A}\}$. This property is expressed by calling the vector Ω_ω a cyclic vector.

(b) For all $X \in \mathfrak{A}$,

$$\omega(X) = (\xi_\omega, \pi_\omega(X)\Omega_\omega)$$

The triplet $\{\pi_\omega, \mathscr{H}_\omega, \Omega_\omega\}$ is called the GNS-representation induced by the state ω.

Moreover if τ is a canonical transformation of $\mathfrak{A}$ leaving the state invariant (i.e. $\omega(\tau(X)) = \omega(X)$ for all X), then there exists a unitary operator U_ω on $\mathscr{H}_\omega$ such that

(i) the cyclic vector is unitary invariant: $U_\omega \Omega_\omega = \Omega_\omega$

(ii) the transformation τ is represented in the GNS-representation by the unitary operator as follows: For each observable X we have $\pi_\omega(\tau(X)) = U_\omega \pi_\omega(X) U_\omega^{-1}$

The operator U_ω is the unitary operator implementing the canonical transformation τ in the GNS-representation of the state ω.

Proof. Note first that the triplet presented in the theorem depends heavily on the given state ω. However for notational convenience the ω-index will be dropped from the notation and therefore, for the proof of this theorem, we construct the triplet denoted simply by $(\mathscr{H}, \pi, \Omega)$. Later we come back to a more detailed discussion about this state dependance together with its physical relevance.

For didactic reasons before treating immediately the general case, we consider first the case of the state being faithful, that is, a state satisfying the property that, if $\omega(X^*X) = 0$ holds for any $X \in \mathfrak{A}$, then $X = 0$ as well. In this case the state defines a scalar product on the (underlying) vector space of $\mathfrak{A}$ as follows: $(X, Y) = \omega(X^*Y)$ (see Eq. (7.1)).

We denote by $\mathscr{H}$ the Hilbert space obtained as the closure of this vector space with respect to this scalar product. Because of this definition of the Hilbert space $\mathscr{H}$.

We now define the representation π by $\pi(Y)X = YX$ for all $X, Y \in \mathfrak{A}$, that is, by left multiplication. Clearly $\pi(Y)$ is densely defined on $\mathscr{H}$ and the observables are represented by bounded operators because

$$||\pi(Y)X||^2 = \omega((YX)^*YX) = \omega(X^*Y^*YX) \le ||Y||^2\omega(X^*X)$$

The continuity Eq. (7.2) defines the operator $\pi(Y)$ everywhere on $\mathscr{H}$. Finally we define the vector Ω by taking $\Omega = \mathbf{1}$, the unit operator, which implies also that $Y = \pi(Y)\Omega$. Note moreover that for all $X, Y, Z \in \mathfrak{A}$,

$$\begin{aligned}
(\pi(X)\pi(Y)\Omega, \pi(Z)\Omega) &= (\pi(Y)\Omega, \pi(X)^*\pi(Z)\Omega) \\
&= \omega(Y^*X^*Z) \\
&= (\pi(Y)\Omega, \pi(X^*)\pi(Z)\Omega)
\end{aligned}$$

or that $\pi(X^*) = \pi(X)^*$, as well as $(\Omega, \pi(X)\Omega) = (\mathbf{1}, \pi(X)\mathbf{1}) = \omega(X)$. This proves the first part, property (a) and (b), of the theorem in the case that the state is faithful.

Now for the hardliners among us, let us relax the condition of faithfulness. Consider the subset J of $\mathfrak{A}$ given by $J = \{X \in \mathfrak{A} \mid \omega(X^*X)\} = 0$.

For all $X \in J$ and $Y \in \mathfrak{A}$,

$$\omega((YX)^*YX) \le \omega(X^*Y^*YX) \le ||Y||^2\omega(X^*X) = 0$$

by the continuity of the state, and for all $X, Y \in J$,

$$\omega((X+Y)^*(X+Y)) \le 2\sqrt{\omega(X^*X)}\sqrt{\omega(Y^*Y)} = 0$$

We conclude that J is a left ideal of $\mathfrak{A}$ as a vector space. Consider the quotient set $\mathfrak{A}/J$ and denote by η_X the rest class with representant X. Define now on $\mathfrak{A}/J$ the scalar product $(\eta_X, \eta_Y) = \omega(X^*Y)$ and again by $\mathscr{H}$ the closure of $\mathfrak{A}/J$ for this scalar product. Define the representation now again by $\pi(Y)\eta_X = \eta_{YX}$ and the cyclic vector by $\Omega = \eta_1$. If we repeat the proof above we will obtain the general proof of the first part of the GNS-theorem. Note also that an easy well-known argument shows the uniqueness of the triplet $(\mathscr{H}, \pi, \Omega)$ up to unitary equivalence. We leave the proof of this property as an exercise.

It remains for us to prove the second part of the theorem dealing with the situation that a state ω is τ-invariant, which is expressed by $\omega = \omega \circ \tau$.

We define the operator U on the dense domain $\{\pi(X)\Omega \mid X \in \mathfrak{A}\}$ of $\mathscr{H}$ as follows: $U\pi(X)\Omega = \pi(\tau(X))\Omega$ for all $X \in \mathfrak{A}$. The operator U is a unitary operator because for all X, Y,

$$(U\pi(X)\Omega, U\pi(Y)\Omega) = \omega(\tau(X^*Y)) = \omega(X^*Y) = (\pi(X)\Omega, \pi(Y)\Omega)$$

or $U^*U = 1$ and as $U\pi(\tau^{-1}(X))\Omega = \pi(X)\Omega$, we get $U^{-1} = U^*$, the unitarity of the operator U. Also

$$\pi(\tau(X))\pi(Y)\Omega = \pi(\tau(X\tau_{-1}(Y)))\Omega = U\pi(X)\pi(\tau^{-1}(Y))\Omega = U\pi(X)U^*\pi(Y)\Omega$$

for all Y and therefore $\pi(\tau(X)) = U\pi(X)U^*$.

Finally we compute for all X

$$\omega(X) = (\Omega, \pi(X)\Omega) = \omega(\tau_{-1}(X)) = (\Omega, U^*\pi(X)\Omega) = (U\Omega, \pi(X)\Omega)$$

Hence $U\Omega = \Omega$. This proves the second part of the theorem.

We should realize that the GNS-construction yields, for any physical system with reversible dynamics, a Hilbert space presentation of the type they learned about in their undergraduate quantum mechanics lectures. With good reasons, we should ask ourselves: What does the GNS-construction add to our knowledge? A partial answer consists in making the observation that the representation space $\mathscr{H}_\omega$, and therefore also the representation π_ω, can be (and also is in general) completely different from the Hilbert space $\mathscr{H} = L^2(\mathbb{R}^d)$, the space of square integrable functions, which is addressed in quantum mechanics books. In particular for boson systems, not all spaces $\mathscr{H}_\omega$ coincide with the Fock space. We can check that only for a highly special class of states ω does the GNS-representation space coincide with this Fock space. This specific kind of question was one of the important topics of the mathematical physics research in field theory and statistical mechanics, which made a considerable progress in the 1960s and 1970s (see [26] for example).

Continuing the discussion on the fact that the GNS-representation triplet $(\mathscr{H}, \pi, \Omega)$ of a state does in general depend greatly on the state, we can add the following: Even if the algebra $\mathfrak{A}$ is a Banach algebra (closed for the norm), weak limit elements of the representant algebra $\pi(\mathfrak{A})$ acting on the Hilbert space $\mathscr{H}$ may depend on the state. A weak limit of a sequence $(A_n)_n$ of elements of the algebra is a limit determined by $\lim_n (\Phi, \pi(A_n)\Psi)$ for any choice Φ, Ψ of $\mathscr{H}$. In the applications of physics many of these limits correspond to the set of highly relevant observable quantities. Indeed for instance we must realize that thermodynamic limits of local operators stand for this kind of weak limit. Therefore in the applications the weak closure of the representees algebra $\pi(\mathfrak{A})$, usually denoted by $\pi(\mathfrak{A})''$, is an important algebra of physically relevant observables which is not universal but very much state dependent and as such very much system dependent.

Also the unitary operator U implementing a canonical transformation in general depends heavily on the state. Consider for instance reversible dynamics $(\tau_t \equiv \alpha_t)_t$ and a stationary state ω. Then the GNS-construction yields for each $t \in \mathbb{R}$ such a unitary operator U_t. We obtain a group of unitary operators $\{U_t | t \in \mathbb{R}\}$ implementing the dynamics within the representation π. In the case it happens that the map $t \to U_t$ is strongly continuous ($\forall \Psi$ in a dense set of $\mathscr{H}$ holds $\lim_{t\to 0} ||(U_t - 1)\Psi|| = 0$), then there exists a self-adjoint operator H such that $U_t = e^{itH}$. The operator H is called the Hamiltonian of the system in the state ω and again heavily depends in general on that state. This is, apart from many other physically interesting situations, also the basis of the notion of generalized "effective Hamiltonian" as it is presented in Chapter Eq. (5) when dealing with the construction of the boson dynamics.

The GNS-construction Eq. (7.1) is performed for a Banach algebra of observables with an involution. The Banach property is essential for the continuity property yielding the result that all representant operators $\pi(X)$ are bounded operators on the representation space $\mathscr{H}$. It is worth considering also the situation of a GNS-construction allowing for representations by unbounded operators. This version of the GNS-construction includes the so-called Wightman-Reconstruction-Theorem [158] in field theory, which is relevant when working with the algebra of observables given by the polynomials of the boson creation and annihilation operators, which are unbounded operators on the boson Fock space. The corresponding version of the GNS-construction is given by the following setup: Suppose we are given a state ω by its set of (truncated) correlation functions $\{c_{n,m}(x,y)_{(t)} \mid n,m \in \mathbb{N}, n+m > 0\}$ with $x = (x_1,...,x_n) \in \mathbb{R}^{nd}$, $y = (y_1,...,y_m) \in \mathbb{R}^{md}$. We suppose that they satisfy some regularity conditions such that they are kernels of tempered distributions and satisfy the necessary positivity conditions for correlation functions of a state on the boson algebra (see Chapter Eq. (2)). This set defines the state ω by the formula: For all $f_i, g_j \in \mathscr{S}$

$$\omega(a^*(f_1)...a^*(f_n)a(g_1)...a(g_m))_{(t)} = \int dxdy \prod_{i=1}^n f_i(x_i) \prod_{j=1}^m g_j(y_j)\, c_{n,m}(x,y)_{(t)}$$

Using a proof along the same lines as the proof of Eq. (7.1), modulo all statements on boundedness, we obtain once again a GNS-triplet $(\mathscr{H}_\omega, \pi_\omega, \Omega_\omega)$ with π_ω a representation on the Hilbert space $\mathscr{H}_\omega$ of the boson canonical commutation relations such that $[\pi_\omega(a(f)), \pi_\omega(a^*(g))] = (f,g)$ and $[\pi_\omega(a(f)), \pi_\omega(a(g))] = 0$. The operators $\pi_\omega(a(f))$ and $\pi_\omega(a^*(g))$ are unbounded operators on $\mathscr{H}_\omega$.

Even more general, take any set of correlation functions defined on any symplectic space (H, σ) satisfying the necessary positivity conditions for the correlation functions of a state on the boson algebra of observables. It defines a state on a boson algebra built on that symplectic space. This algebra is denoted by $W(H, \sigma)$ as defined in Eq. (6.9). The proof of the GNS-construction yields again a representation of this boson algebra on the canonical Hilbert space constructed in Eq. (7.1). In particular this scheme is the technology which is applied in the theory of quantum fluctuations as explained in Chapter Eq. (6).

7.2 Dynamical Semigroups

As announced above we also consider, next to the dynamical groups describing reversible systems, the dynamical semigroups $(\alpha_t)_{t\in\mathbb{R}^+}$, describing irreversible systems. This subject is a vast, diverse, and a growing topic which is however still very much in a state of development. Therefore we consider only a special class of these irreversible dynamics, namely those which are called of the Markovian type and which are of the form $\alpha_t = \exp(tL)$, where L is a linear operator densely defined and acting on and into the algebra of observables. In this situation, the operator L is called the *generator of the dynamics*. To keep the mathematics simple and to explain the relation between the properties of such a dynamic and the structural properties of its generator L, we start with an operator L which we assume to be bounded with norm $||L|| = \sup_{X\in\mathfrak{A}} ||L(X)|| < \infty$ acting on the normed algebra of observables $\mathfrak{A}$.

We consider the linear map L which satisfies the following three properties:

(i) *unity conserving*, $L(1) = 0$,

(ii) *self-adjoint*, $L(X^*) = L(X)^*$ for all observables X and

(iii) *dissipative*, that is, satisfying for each observable X the inequality

$$L(X^*X) \geq X^*L(X) + L(X^*)X \tag{7.3}$$

Each map L with these three properties is called a *dissipative generator*. indexdissipative generator

We are already familiar with the notion of *derivation*. This is a linear operator L satisfying the three properties above, in particular also the property Eq. (7.3), supplemented with the equality $L(X^*X) = X^*L(X) + L(X^*)X$ instead of the inequality sign. Of course the best known example of such a derivation is given by the map $L = i[H, .]$, where $H^* = H$ is any self-adjoint operator. In this case we get for $\alpha_t = \exp(it[H, .])$ a conservative or reversible dynamic treated in the previous section. Modulo some mathematical technicalities it is easy to understand that L is a derivation if and only if the $(\alpha_t \mid t \in \mathbb{R})$ form a one-parameter group of *-automorphisms or of canonical transformations of the algebra of observables ([46] Theorem 14.1).

Again for simplicity of the mathematical argumentation, take as algebra of observables $\mathfrak{A}$, a subalgebra of some $\mathbb{B}(\mathscr{H})$, the set of bounded operators acting on a Hilbert space $\mathscr{H}$. In this case a minimal knowledge of operator theory (like some spectral theory), leads already to an easily understandable proof of the following theorem:

Theorem 7.2. *The map L can be exponentiated for all $t \geq 0$, that is, the exponential e^{tL} is well defined for all $t \geq 0$ and is norm continuous in t. Furthermore the following properties are equivalent: For all $X \in \mathfrak{A}$,*

1. *e^{tL} is a positive map for all $t \geq 0$; in particular, this means that for all observables X,*

$$e^{tL}(X^*X) \geq e^{tL}(X^*)e^{tL}(X) \qquad (7.4)$$

   and the semigroup of maps $\{e^{tL} \mid t \in \mathbb{R}^+\}$ is called a dissipative dynamics.
2. *L is dissipative generator, in particular it satisfies*

$$L(X^*X) \geq X^*L(X) + L(X^*)X \qquad (7.5)$$

Proof. Because the operator L is supposed to be norm bounded the exponentiation property, using the series expansion of the exponential function and the norm convergence of this function, the existence of $\exp(tL)$ for each t, is immediate. The norm continuity in t is also immediate.

Furthermore, we find that item 1 implies item 2 by differentiating the first inequality with respect to the t-variable at $t = 0$.

We now consider the inverse implication, that is, item 2 implies item 1.

Let us first prove that e^{tL} is a positive map, that is, a mapping of positive operators into positive operators, for all $t \geq 0$. We use the well-known formula $e^{tL} = \lim_n (1 - \frac{tL}{n})^n$. It follows from this expression that it is sufficient to prove that $(\lambda - L)^{-1}$ is positive for all sufficiently large positive λ. Next we consider $\lambda > ||L||$. To show that $(\lambda - L)^{-1} \geq 0$, it is enough to show that $X \geq 0$ whenever X is self-adjoint and satisfies $(\lambda - L)X \geq 0$. We prove this as follows. Let $X = X^* = X_+ - X_-$, with $X_\pm$ representing the positive and negative parts of the operator X satisfying $X_- X_+ = 0$ and both $X_\pm \geq 0$. We now show that $X_- = 0$. From $(\lambda - L)X \geq 0$,

$$0 \leq X_-(\lambda - L)(X)X_- = -X_-^3\lambda + X_-L(X_-)X_- - X_-L(X_+)X_-$$

Note that X_+ is necessarily of the form $X_+ = Y^2$ for some Y with the properties $Y^* = Y$, $YX_- = 0$. Hence from item 2 it follows that $X_-L(X_+)X_- \geq 0$. Therefore

$$\lambda X_-^3 \leq X_-L(X_-)X_-$$

Taking the norm of this inequality yields $\lambda ||X_-||^3 \leq ||X_-||^3||L||$. As $||L|| < \lambda$, we get $||X_-|| = 0$ or $X_- = 0$. This proves the positivity of the exponential map e^{tL} for all $t \geq 0$.

We now define the function $f(t)$

$$f(t) = e^{tL}(X^*X) - e^{tL}(X^*)e^{tL}(X)$$

Taking the derivative with respect to t we get

$$f'(t) = L(e^{tL}(X^*X)) - L(e^{tL}(X^*))e^{tL}(X) - e^{tL}(X^*)L(e^{tL}(X))$$

such that

$$
\begin{aligned}
f(t) - e^{tL}f(0) &= \int_0^t \frac{d}{ds}[e^{(t-s)L}f(s)]ds \\
&= -\int_0^t e^{(t-s)L}Lf(s)ds + \int_0^t e^{(t-s)L}\frac{d}{ds}f(s)\,ds \\
&= \int_0^t e^{(t-s)L}\{L(e^{sL}(X^*)e^{sL}(X)) \\
&\quad -(Le^{sL}(X^*))e^{sL}(X) - e^{sL}(X^*)(Le^{sL}(X))\}\,ds
\end{aligned}
$$

The expression between the large brackets is positive by hypothesis. Moreover $e^{(t-s)L}$ is positive for $0 \leq s \leq t$, hence $f(t) \geq e^{tL}f(0) = 0$. This proves item 1, completing the proof of the theorem.

This theorem shows that the dynamics $\alpha_t = \exp(tL), t \geq 0$ is well defined as an irreversible or dissipative dynamics. It is proven to exist if the generator L satisfies the conditions of the theorem, particularly if the map L is a bounded map. In the proof the norm topology is used throughout.

For practical applications of this result to boson systems, we should have noted that the boundedness condition poses an immediate mathematical technical problem. However as is clear from Chapter Eq. (5), the definition of the reversible as well as the irreversible dynamics are formulated in terms of the weak operator topologies, that is, under some given state or for a particular set of states—see Eq. (5.3). It is outside the scope of this discussion to enter into all the mathematical details of the proofs about the equivalence between points 1 and 2 of the equivalent theorem working with unbounded operators in these weak topologies. To get an idea of such proofs in the case of unbounded operators in the literature, we can refer to [161, 45]. Furthermore, in Chapter Eq. (5) we find model computations with unbounded operators which are completely worked out.

In any case if we can apply the theorem and obtain the dynamics $\alpha_t = \exp(tL)$, then it is clear that for each observable X we have a dynamical equation of the type

$$\frac{dX_t}{dt} = LX_t$$

for the quantities $X_t = \alpha_t X$. This is again the Heisenberg picture of what is called in physics the *master equation* [36]. As in the reversible case, we can as well consider its Schrödinger picture.

We should stress that the type of dynamics discussed above is a very special type of dynamical semigroup in the sense that they have not only the property of positivity but also the mathematical property of *complete positivity*. We will not dive into all the details of this stronger notion of positivity. Instead we refer again to the literature for the most physics-minded results about complete positivity, as well as for the history of the topic about irreversible dynamics [2, 46]. Earlier work on the generators of dynamical semigroups and the dissipative evolutions started with the work of Kossakowski and Ingarden [91, 92, 93, 84]. Basic references are also [36, 107].

As a last remark about the systems equipped with a dynamic sharing the properties of this special class of positive mappings, namely those of the type of being completely positive, they share the interesting property of being dilationable quantum mechanical dynamical system. This means that each such dynamical system, equipped with such a completely positive dissipative dynamics, can always be embedded into a larger system endowed with a reversible or conservative dynamic. Moreover the original irreversible dynamics are the restriction to the original smaller system of the larger reversible system. In other words this mathematical observation of complete positivity provides us with a good argument of the fact that a completely positive dynamic is always the restriction to a subsystem of a larger conservative system. The larger system comes across as a composed system, as the union of the small system coupled to a rest system. Therefore the latter one is endowed with the interpretation of being a heat bath for the small system [46, 102].

More information about solvable irreversible dynamics relevant for the applications in boson systems can also be found in [38, 161].

7.3 Canonical Transformations

In the study of boson systems the notion of canonical transformation has always played an important role. Also in this monograph it enters the discussion at several places. Several different types of canonical transformations are used. In order to see better their differences and their similarities they are brought together in this section.

A *canonical transformation of the boson algebra of observables* $\mathfrak{A}$, acting on the Fock space representation or on any other of its representations, is a one-to-one *-algebra morphism of $\mathfrak{A}$ onto itself leaving the canonical commutation relations invariant.

Mathematically, this means that a canonical transformation is a map, say τ, of $\mathfrak{A} \to \mathfrak{A}$ which has an inverse τ^{-1}, and which is a *-algebra morphism* which means that it has the properties satisfying

1. linearity: $\forall \lambda, \mu \in \mathbb{C}$ and $X, Y \in \mathfrak{A}: \tau(\lambda X + \mu Y) = \lambda \tau(X) + \mu \tau(Y)$
2. *-invariance : $\tau(X^*) = \tau(X)^*$
3. product conserving $\tau(XY) = \tau(X)\tau(Y)$

and which leaves the canonical commutation relations invariant, that is, for all $f, g \in \mathscr{S}$,

$$[\tau(a(f)), \tau(a^*(g))] = (f, g)\mathbf{1}; \quad [\tau(a(f)), \tau(a(g))] = 0 \qquad (7.6)$$

The following types of canonical transformations are explicitly used in this monograph:

1. Let U be any unitary operator acting on a Fock space, or on any other representation space of a CCR-algebra. Define the map τ on $\mathfrak{A}$ by $\tau(X) = UXU^*$ as a *-algebra morphism as defined above. We can easily check that the map τ satisfies Eq. (7.6) and therefore is a canonical transformation. We can say that this canonical transformation τ is *implemented* by the unitary operator U.
   Immediate examples of this type of canonical transformations are the following:
   – Consider the special group of reversible time evolutions Eq. (5.1) $\{\alpha_t | t \in \mathbb{R}\}$ of ordinary quantum mechanics. For each real value of the time t and for each self-adjoint operator H, called a Hamiltonian, we consider the unitary operator $U_t = e^{itH}$. Clearly $\alpha_t(X) = U_t X U_{-t}$ is such a canonical transformation.
   – The group of gauge transformations Eq. (2.28) $\{\tau_\lambda | \lambda \in [0, 2\pi]\}$. Consider any representation of the CCR-algebra for which $N = \int dx\, a^*(x)a(x)$ is the self-adjoint number operator. Consider the unitary operators for any real λ: $U_\lambda = e^{i\lambda N}$. We compute $\tau_\lambda(a(f)) = U_\lambda a(f)U_{-\lambda} = e^{-i\lambda}a(f)$ and hence each τ_λ is again a canonical transformation.
2. Note that the gauge transformations can also be defined directly, without using the number operator, by $\tau_\lambda(a(f)) = e^{-i\lambda}a(f)$ and by extension as *-algebra morphism to the whole algebra. The number operator is not used in the definition. The gauge transformations satisfy also Eq. (7.6) and are therefore canonical transformations for all real λ. This situation occurs when working in representations in which the number operator cannot be given a meaning. In a more physical language, this happens for instance in representations of the boson algebra with a cyclic vector carrying an infinity of particles. Such systems might carry an infinite number of particles but a finite density of particles. This is the situation typical for equilibrium states of boson systems with a non-zero density of particles. A similar remark can be made for the reversible time evolutions with a possible non-existence of a well-defined Hamiltonian. This might again be the case of thermodynamic limit systems carrying an infinite energy but a finite energy density.
3. Bose field translations Eq. (2.15).
   Let χ be any real linear functional on the test function space $\mathscr{S}$. Define the boson field transformation $\tau_\chi(b(f)) = b(f) + \chi(f)$ and extend it to the whole algebra as a *-algebra morphism. It satisfies Eq. (7.6) and is therefore a canonical transformation.
4. The group of space translations and any of its subgroups.

Let $\tau_x(a(f)) = a(f_x)$, where $f_x(y) = f(y-x)$ is the translated function over the distance $x \in \mathbb{R}^d$. Extend the transformations to the whole algebra by means of the *-algebra morphism properties. All maps $\{\tau_x \,|\, x \in \mathbb{R}^d\}$ satisfy the criterion Eq. (7.6) as a consequence of the property that the scalar product is invariant under the translations, in particular $(f_x, g_x) = (f, g)$ for all x. All elements of the group $\{\tau_x \,|\, x \in \mathbb{R}^d\}$ are canonical transformations.

5. The group of space rotations and any of its subgroups.

   Define $\tau_R(a(f)) = a(f_R)$ for any R, element of $O(d)$, the group of orthogonal rotations in d dimensions, hence satisfying the property $R^+R = 1$, and where $f_R(x) = f(R^{-1}x)$. Extend again these maps τ_R as *-algebra morphisms to the whole CCR-algebra of observables. As a consequence of the rotation invariance of the scalar product $(f, g) = (f_R, g_R)$, all τ_R satisfy Eq. (7.6) and are therefore canonical transformations. We obtain a group of canonical transformations isomorphic with the rotations group $O(d)$. We can proceed analogously for any of the subgroups of the full rotation group $O(d)$.

6. Bogoliubov transformations Eq. (2.30).

   For each $p \in \mathbb{R}^d$ consider the *-algebra morphism γ_p of $\mathfrak{A}$ defined by

$$\gamma_p(a(p)) \equiv \tilde{a}(p) = u(p)a(p) - v(p)a^*(-p)$$

where u, v are real functions on $\mathbb{R}^d$ satisfying the properties $u(-p) = u(p)$, $v(-p) = v(p)$ and $u(p)^2 - v(p)^2 = 1$ for each choice of p. We extend the γ_p as a *-algebra morphism to the whole algebra generated by the creation and annihilation operators. We check that each γ_p satisfies the conditions Eq. (7.6) and hence that all the γ_p are canonical transformations which are called Bogoliubov transformations.

7. Condensate canonical transformation Eq. (4.18).

   Let U be any unitary operator commuting with all creation and annihilation operators acting on any representation space of the canonical boson commutation relations. We define the *-algebra morphism η of $\mathfrak{A}$ by $\eta(a(f)) = U^*a(f)$. We check again that it satisfies Eq. (7.6) and hence that it is a canonical transformation. This type of canonical transformation can also be seen as a generalization of special cases considered in the first item (e.g., the gauge transformations defined without the use of the number operator). These canonical transformations were of great use in our analysis of spontaneous symmetry breaking Eq. (4.3.2).

References

1. Accardi L., Bach A.: *Quantum central limit theorems for strongly mixing random variables*, Zeitschrift Wahrscheinlichkeitsth. Verw. Geb. **68**, 393–402 (1985)
2. Alicki R., Fannes M.: *Quantum dynamical sytems*, Oxford University Press (2001)
3. Alicki R., Lendi K.: *Quantum dynamical semigroups and applications*, Lecture Notes in Physics, Springer Verlag, Berlin-Heidelberg, 2007
4. Anderson P.W.: *Random-Phase Approximation in the Theory of Superconductivity*, Phys. Rev. **112**, 1900–1916 (1958)
5. Anderson P.W.: *Two crucial experimental tests of the resonating valence bond-Luttinger liquid interlayer tunneling theory of high-Tc superconductivity*, Phys. Rev. B **64**, 2624–2626 (1990)
6. Anderson M.H., Ensher J.R., Matthews M.R., Wiemann C.E., Cornell E.A.: *Observation of Bose-Einstein condensation in a dilute atomic vapor*, Science **269**, 198–201 (1995)
7. Angelescu N., Verbeure A.: *Variational solution of a superfluidity model*, Physica A **216**, 386–396 (1995)
8. Angelescu N., Verbeure A., Zagrebnov V.A.: *On Bogoliubov's model of superfluidity*, J. Phys. A **25**, 3473–3491 (1992)
9. Angelescu N., Verbeure A., Zagrebnov V.A.: *Superfluidity III*, J. Phys. A **30**, 4895–4913 (1997)
10. Angelescu N., Verbeure A., Zagrebnov V.A.: *Quantum n-vector anharmonic crystal I: $1/n$-expansion*, Commun. Math. Phys. **205**, 81–95 (1999)
11. Angelescu N., Verbeure A., Zagrebnov V.A.: *Quantum n-vector anharmonic crystal II: displacement fluctuations*, J. Stat. Phys. **100**, 829–851 (2000)
12. Araki H., Woods E.J.: *Representations of the canonical commutation relations describing a non-relativistic infinite free Bose gas*, J. Math. Phys. **4**, 637–672 (1963)
13. Beliaev S.T.: *Energy spectrum of a non-ideal Bose gas*, Sov. Phys.-JEPT **7**, 299–307 (1958)
14. van den Berg M.: *On Bose condensation into an infinite number of low-lying levels*, J. Math. Phys. **23**, 1159–1161 (1982)
15. van den Berg M., Lewis J.T.: *On the free Bose gas in a weak external field*, Commun. Math. Phys. **81**, 475–494 (1981)
16. van den Berg M., Lewis J.T.: *On generalised condensation in the free Bose gas*, Physica A **110**, 550–564 (1982)
17. van den Berg M., Lewis J.T., Lunn M.: *On the general theory of Bose-Einstein condensation and the state of the free Bose gas*, Helv. Phys. Acta **59**, 1289–1310 (1986)

A.F. Verbeure, *Many-Body Boson Systems*, Theoretical and Mathematical Physics,   179
DOI 10.1007/978-0-85729-109-7, © Springer-Verlag London Limited 2011

18. van den Berg M., Lewis J.T., Pulé J.V.: *A general theory of Bose-Einstein condensation*, Helv. Phys. Acta **59**, 1271–1288 (1986)

19. Binney J.T., Dowrick N.J., Fisher A.J., Newman M.E.J.: *The theory of critical phenomena*, Clarendon, Oxford, 1992

20. Bogoliubov N.N.: *On the theory of superfluidity*, J. Phys. (USSR) **11**, 23–32 (1947)

21. Bogoliubov N.N.: *About the theory of superfluidity*, Izv. Akad. Nauk. (USSR) **11**, 77–90 (1947)

22. Bogoliubov N.N.: *Energy levels of the imperfect Bose-Einstein gas*, Bull. Moscow State Univ. **7**, 43–56 (1947)

23. Bogoliubov N.N.: *Lectures on quantum statistics, Vol 2*, 1970, Gordon and Breach, London

24. Bose S.N.: *Planck's Gezetz und Lichtquantenhypothese*, Z. Phys. **26**, 178–181 (1924)

25. Bouziane M., Martin P.A.: *Bogoliubov inequality for unbounded operators and Bose gas*, J. Math. Phys. **17**, 1848–1851 (1976)

26. Bratteli O., Robinson D.W.: *Operator algebras and statistical mechanics I, II*, (I) 1979, (II, 2nd Edition) 1996, Springer-Verlag, Berlin Heidelberg New York

27. Broidioi M., Momont B., Verbeure A.: *Lie algebra of anomously scaled fluctuations*, J. Math. Phys. **36**, 6746–6757 (1995)

28. Broidioi M., Nachtergaele B., Verbeure A.: *The Overhauser model*, J. Math. Phys. **32**, 2929–2935 (1991)

29. Broidioi M., Verbeure A.: *Plasmon frequency for a spin-density wave model*, Helv. Phys. Acta **64**, 1093–1112 (1991)

30. Broidioi M., Verbeure A.: *The plasmon in the one-component plasma*, Helv. Phys. Acta **66**, 155–180 (1993)

31. Buffet E., de Smedt Ph., Pulé J.V.: *The condensate equation for some Bose systems*, J. Phys. A: Math. Gen., **16**, 4307–4324 (1983)

32. Buffet E., Pulé J.V.: *Fluctuation properties of the imperfect Bose gas*, J. Math. Phys. **24**, 1608–1616 (1983)

33. Cushen C.D., Hudson R.L.: *A quantum central limit theorem*, J. Appl. Probab. **8**, 454–469 (1971)

34. Dalfovo F., Giorgini S., Pitaevskii L.P., Stringari S.: *Theory of Bose-Einstein condensation in trapped gases*, Rev. Mod. Phys. **71**, 463–512 (1999)

35. Davies E.B.: *Quantum theory of open sytems*, Academic Press, London (1976)

36. Davidov A.: *Théorie du solide*, Editions MIR, 1980

37. Davis K.B., Mewes M.O., Andrews M.R., van Druten N.J., Durfee D.S., Kurn D.M., Ketterlee W.: *Bose-Einstein condensation in a gas of sodium atoms*, Phys. Rev. Lett. **75**, 3969–3973 (1995)

38. Demoen B., Vanheuverzwijn P., Verbeure A.: *Completely positive maps on the CCR-algebra*, Lett. Math. Phys. **2**, 161–166 (1977)

39. De Roeck W., Maes C., Netocký K.: *H-Theorems from Macroscopic Autonomous Equations*, J. Stat. Phys. **123**, 571–584 (2006)

40. Dubin D.A., Hennings M.A., Smith, T.B.: *Mathematical Aspects of Weyl Quantization and Phase*, World Scientific Press (Singapore), 2000

41. Dunford N., Schwartz J.T.: *Linear operators, Part II*, Interscience Publishers, NY-London, 1963

42. Einstein A.: *Quantentheorie des einatomigen idealen gases*, Sitzungber. Preuss. **9**, 3–14 (1925)

43. Elgart A., Erdös L., Schlein B., Yau H.T.: *Gross-Pitaevskii equation as the mean field limit of weakly coupled bosons*, Arch. Rational Mech. Anal. **179**, 265–283 (2006)

44. Esposito R., Pulvirenti M.: *From particles to fluids*, Contribution to Handbook of Mathematical Fluid Dynamics, 3 D. Serre and F. Friedlander Editors, Elsevier (2004)

45. Evans D.E.: *Irreducible quantum dynamical semigroups*, Commun. Math. Phys. **54**, 293–297 (1977)

46. Evans D.E., Lewis J.T.: *Dilations of irreversible evolutions in algebraic quantum theory*, Communications of the Dublin Institute for Advanced Studies, Series A (Theoretical Phyics), No 24 (1977)

47. Fannes M.: *The entropy density of quasi-free states for a continuous boson system*, Ann. Inst. Henri Poincaré **28**, 187–196 (1978)

48. Fannes M., Pulé J.V., Verbeure A.: *Goldstone's theorem for bose systems*, Lett. Math. Phys. **6**, 385–389 (1982)

49. Fannes M., Pulé J.V., Verbeure A.: *On Bose condensation*, Helv. Phys. Acta **55**, 391–399 (1982)

50. Fannes M., Spohn H., Verbeure A.: *Equilibrium states for mean fields*, J. Math. Phys. **19**, 355–358 (1980)

51. Fannes M., Verbeure A.: *On the time evolution automorphisms of the CRR-algebra for quantum mechanics*, Commun. Math. Phys. **35**, 257–264 (1974)

52. Fannes M., Verbeure A.: *Correlation inequalities and equilibrium states I and II*, Commun. Math. Phys. **55**, 125–131 (1977); **57**, 165–171 (1977)

53. Fannes M., Verbeure A.: *Global thermodynamical stability and correlation inequalities*, J. Math. Phys. **19**, 558–560 (1978)

54. Fannes M., Verbeure A.: *The condensed phase of the imperfect Bose gas*, J. Math. Phys. **21**, 1809–1818 (1980)

55. Fannes M., Vanheuverzwijn P., Verbeure A.: *Energy entropy inequalities for classical lattice systems*, J. Stat. Phys. **29**, 547–560 (1982).

56. Fannes M., Werner R.F.: *Boundary conditions for quantum lattice systems*, Helv. Phys. Acta **68**, 635–657 (1995)

57. Ginibre J.: *On the asymptotic exactness of the Bogoliubov approximation for many boson systems*, Commun. Math. Phys. **8**, 26–51 (1968)

58. Girardeau M., Arnowitt R.: *Theory of many boson sytems, Pair theory*, Phys. Rev. **113**, 755–761 (1959)

59. Girardeau M.: *Variational method for the quantum statistics of interacting particles*, J. Math. Phys. **3**,131–139 (1962)

60. Giri N., von Waldenfelds W.: *An algebraic version of the central limit theorem*, Zeitschrift Wahrscheinlichkeitsth. Verw. Geb. **42**, 129–134 (1978)

61. Goderis D., Verbeure A., Vets P.: *Non-commutative central limits*, Prob. Th. Rel. Fields **82**, 527–544 (1989)

62. Goderis D., Verbeure A., Vets P.: *Theory of quantum fluctuations and the Onsager relations*, J. Stat. Phys. **56**, 721–746 (1989)

63. Goderis D., Verbeure A., Vets P.: *Dynamics of fluctuations for quantum lattice systems*, Commun. Math. Phys. **128**, 533–549 (1990)

64. Goderis D., Verbeure A., Vets P.: *Glauber dynamics of fluctuations*, J. Stat. Phys. **62**, 759–771 (1991)

65. Goderis D., Verbeure A., Vets P.: *About the Exactness of the Linear Response Theory*, Commun. Math. Phys. **136**, 265–283 (1991)

66. Goderis D., Vets P.: *Central limit theorem for mixing quantum systems and the CCR-algebra of fluctuations*, Commun. Math. Phys. **122**, 249–265 (1989)

67. Goldstein B.G.: *A central limit theorem of non-commutative probability theory*, Theory of Prob. and its Appl. **27**, 703 (1982)

68. Goldstone J.: *Field theories with superconductor solutions*, Nuovo Cimento **19**, 154–164 (1961)

69. Griffin A., Snoke D.W., Stringari S.: *Bose-Einstein Condensation*, Univ. Press, Cambridge, 1996

70. de Groot S.R., Masur P.: *Non-equilibrium thermodynamics*, North-Holland Publ. Comp. (1962)

71. Gross E.P.: *Structure of a quantized vortex in boson systems*, Il Nuovo Cimento **20**, 454–466 (1961)

72. Gross E.P.: *Hydrodynamics of a superfluid condensate*, J. Math. Phys. **4**, 195–207 (1963)

73. Haag R., Hugenholz N.M., Winnink M.; *On the equilibrium states in quantum statistical mechanics*, Commun. Math. Phys. **5**, 215–236 (1967)

74. Haag R., Kadison R.V., Kastler D.: *Asymptotic orbits in a free Fermi gas*, Commun. Math. Phys. **33**, 1–22 (1973)

75. Heidenreich R., Seiler R., Uhlenbrock D.A.: *The Luttinger model*, J. Stat. Phys. **22**, 27–57 (1980)

76. Hein S., Roepstorff G.: *Vortices in infinite free Bose systems*, Annal. Inst. Henri Poincaré **32**, 21–31 (1980)

77. Hohenberg P.C.: *Existence of long-range order in one and two dimensions*, Phys. Rev. **158**, 383–386 (1967)

78. Hörmander L.: *Estimates for translation invariant operators in L_p-spaces*, Acta Mathematica **104**, 93–140 (1960)

79. Hugenholz N.M.: *Ground state of a system of interacting bosons*, Physica **26**, 170–173 (1960)

80. Hugenholz N.M.: *Quantum theory of many-body systems*, Rep. Prog. Phys. **28**, 201–247 (1965)

81. Hugenholz N.M., Pines D.: *Ground state energy and excitation spectrum of system of interacting bosons*, Phys. Rev. **116**, 489–506 (1959)

82. Iadonisi G., Marinaro M., Vsudevan R.: *Possibility of two stages of phase transition in an interacting Bose gas*, Il Nuovo Cimento **LXX B**, 147–164 (1970)

83. Ibragimov I.A., Linnick Y.V.: *Independent and stationary sequences of random variables*, Wolters-Noordhoff (1971)

84. Ingarden R., Kossakowski A.: *On the connection of non-equilibrium information thermodynamics with non-hamiltonian quantum mechanics of open systems*, Ann. Phys. **89**, 451–485 (1975)

85. Inönü E., Wigner E.P.: *Representations of the Galilei group*, Il Nuovo Cimento **9**, 705–718 (1952)

86. Inönü E., Wigner E.P.: *On the contraction groups and their representations*, Proc. Nat. Acad. Sci. (US) **39**, 510–524 (1953)

87. Jakšić V., Pautrat Y., Pillet C.-A.: *Central limit theorem for locally interacting Fermi gas*, Commun. Math. Phys. **285**, 175–217 (2009)

88. Ketterle W., Inouye S.: *Does matter amplification works for fermions*, Phys. Rev. Lett. **89**, 4203–4206 (2001)

89. Kittel C.: *Quantum theory of solids*, Wiley, 1963

90. Kobe D.H.: *Single particle condensate and pair theory of a homogeneous Bose system*, Ann. Phys. **47**, 15–39 (1968)

91. Kossakowski A.: *On quantum statistical mechanics of non-hamiltonian systems*, Rep. Math. Phys. **3**, 247–274 (1972)

92. Kossakowski A.: *On the necessary and suffcient conditions for a generator of a quantum dynamical semigroup*, Bull. Acad. Polon. Sér. Sci. Math. Astonom. Phys. **20**, 1021–1025 (1972)

93. Kossakowski A.: *On the general form of a generator of a dynamical semigroup for the spin 1/2 system*, Bull. Acad. Polon. Sér. Sci. Math. Astonom. Phys. **21**, 649–653 (1973)

94. Kubo R., Toda M., Hashitsume N.: *Statistical Physics II, Non-equilibrium statistical mechanics*, Springer, Berlin-Heidelberg-New York, 1985

95. Landau L.D.: *The theory of superfluidity of Helium II*, J. Phys. (USSR) **5**, 71 (1941)

96. Landau L.D.: *On the theory of superfluidity of Helium II*, J. Phys. (USSR) **11**, 91 (1947)

97. Lanford D.E., Ruelle D.: *Observables at infinity and states with short range correlations in statistical mechanics*, J. Math. Phys. **13**, 194–215 (1969)

98. Lauwers J., Verbeure A.: *Fluctuations in the Bose gas with attractive boundary conditions*, J. Stat. Phys. **108**, 123–168 (2002)

99. Lauwers J., Verbeure A., Zagrebnov V.A.: *Bose-Einstein condensation for homogeneous interacting sytems with a one-particle spectral gap*, J. Stat. Phys. **112**, 397–420 (2003)

100. Lebowitz J., Spohn H.: *Irreversible thermodynamics for quantum sytems weakly coupled to thermal reservoirs*, Adv. Chem. Phys. **39**, 109–142 (1978)

101. Lewis J.T., Pulé J.V.: *The equilibrium states of the free Bose gas*, Commun. Math. Phys. **36**, 1–18 (1974)

102. Lewis J.T., Thomas L.C.: *How to make a heat bath*, Functional integration, ed A.M. Arthurs, Proc. Int. Conf., Cumberland Lodge, London 1974; Oxford, Clarendon Press

103. Lieb E.H., Mattis D.C.: *Mathematical physics in one dimension*, Academic, New York (1966)

104. Lieb E.H., Seiringer R.: *Proof of Bose-Einstein condensation for dilute trapped gases*, Phys. Rev. Lett. **88**, 170409 (2002)

105. Lieb E.H., Seiringer R., Yngvason J.: *Bosons in a trap: a rigorous derivation of the Gross-Pitaevskii energy functional*, Phys. Rev. A 61, 043602-1–043602-13 (2000)

106. Lieb E.H., Seiringer S., Yngvason J.: *A rigorous derivation of the Gross-Pitaevskii energy functional for a two-dimensional Bose gas*, Commun. Math. Phys. 224, 17–31 (2001)

107. Lindblad G.: *On the generators of quantum dynamical semigroups*, Commun. Math. Phys. **48**, 119–130 (1976)

108. London F.: *On the Bose-Einstein condensation*, Phys. Rev. **54**, 947–954 (1938)

109. London F.: *Superfluids, Vol. II*, Wiley, New York (1954)

110. Luban M.: *Statistical mechanics of a non-ideal boson gas*, Phys. Rev. **128**, 965–987 (1962)

111. Luttinger J.M.: *An exactly soluble model of a many-fermion system*, J. Math. Phys. **4**, 1154 (1963)

112. Maes C.: *On the origin and the use of fluctuation relations for the entropy*, Séminaire Poincaré **2**, 29–62 (2003)

113. Maes C., Tasaki Hal: *Second law of thermodynamics for macroscopic mechanics coupled to thermodynamic degrees of freedom*, Lett. Math. Phys. **79**, 251–261 (2007)

114. Maes C., Netockny K., Shergelashvili B.: *A selection of nonequilibrium issues*, Lecture notes from the 5th Prague Summer School on Mathematical Statistical Mechanics (2006)

115. Manuceau J., Verbeure A.: *Quasi-free states of the CCR-algebra and Bogoliubov ttransformations*, Commun. Math. Phys. **9**, 293–302 (1968)

116. Martin P.A.: *A remark on the Goldstone Theorem in statistical mechanics*, Nuovo Cimento **68B**, 302–313 (1982)

117. Martin Ph., Rothen F.: *Problèmes à N-corps et champs quantiques*, Presse Ecole Polytechnique et Univ. Romandes, Lausanne, 1990
118. Matsui T.: *Bosonic central limit theorem for one-dimensional XY-model*, Rev. Math. Phys. **14**, 675–700 (2002)
119. Matsui T.: *On the algebra of fluctuations in quantum spin chains*, Ann. Henri Poincaré **4**, 63–83 (2003)
120. Mermin N., Wagner H.: *Absence of ferromagnetism or anti-ferromagnetism in one- or two-dimensional isotropic Heisenberg models*, Phys. Rev. Lett. **17**, 1133–1136 (1966)
121. Messer J., Verbeure A.: *Free bosons in a scaled external potential*, J. Phys. A 15, L111–L114 (1982)
122. Michoel T., Momont B., Verbeure A.: *CCR-structure of normal k-mode fluctuations*, Reports Math. Phys. **41**, 361–395 (1998)
123. Michoel T., Verbeure A.F.: *Non-extensive Bose-Einstein condensation model*, J. Math. Phys. **40**, 1268–1279 (1999)
124. Michoel T., Verbeure A.: *Goldstone boson normal coordinates in interacting Bose gases*, J. Stat. Phys. **96**, 1125–1162 (1999)
125. Michoel T., Verbeure A.: *Goldstone boson normal coordinates*, Commun. Math. Phys. **216**, 461–490 (2001)
126. Momont B., Verbeure A.: *Algebraic structure of k-mode fluctuations*, Reports Math. Phys. **40**, 97–105 (1997)
127. Momont B., Verbeure A., Zagrebnov V.A.: *Algebraic structure of quantum fluctuations*, J. Stat. Phys. **89**, 633–653 (1997)
128. Naudts J., Verbeure A., Weder R.: *Linear response theory and the KMS-condition*, Comm. Math. Phys. **44**, 87–100 (1975)
129. Penrose O., Onsager L.: *Bose-Einstein condensation and liquid helium*, Phys. Rev. **104**, 576 (1956)
130. Perkins A.S., Walls D.F.: *The physics of trapped dilute gas Bose-Einstein condensates*, Phys. Rep. **303**, 1–80 (1998)
131. Petz D.: *An invitation to the algebra of canonical commutation relations*, Leuven Notes in Math. and Theor. Phys. Vol. 2 (1990), Leuven University Press
132. Pines D.: *Elementary excitations in solids*, Benjamin, New York, 1964
133. Pines D., Nozières P.: *The theory of quantum liquids*, Benjamin, New York, 1966
134. Pitaevskii L.P.: *Vortex lines in an imperfect Bose gas*, Sov. Phys. JEPT, **13**, 451–454 (1961)
135. Pitaevskii L., Stringari S.: *Bose-Einstein Condensation*, Clarendon Press, Oxford, 2003
136. Pulé J.V.: *The free Bose gas in a weak external field*, J. Math. Phys. **24**, 138–142 (1983)
137. Pulé J.V., Verbeure A.F., Zagrebnov V.A.: *Models for equilibrium BEC superradiance*, J. Phys. A: Math. Gen. **37**, L321–L328 (2004)
138. Pulé J.V., Verbeure A.F., Zagrebnov V.A.: *On non-homogeneous Bose condensation*, J. Math. Phys. **46 (8)**, Art. No 083301 (2005)
139. Pulé J.V., Verbeure A.F., Zagrebnov V.A.: *A Dicke type model for equilibrium BEC superradiance*, J. Stat. Phys. **119**, 309–329 (2005)
140. Pulé J.V., Verbeure A.F., Zagrebnov V.A.: *Models with recoil for BEC and superradiance*, J. Phys. A: Math. Gen. **38**, 5173–5192 (2005)
141. Pulé J.V., Verbeure A.F., Zagrebnov V.A.: *On solvable Boson Models*, J. Math. Phys. **49**, 043302(18p) (2008)
142. Pulé J.V., Zagrebnov V.A.: *A pair Hamiltonian of a nonideal boson gas*, Ann. Inst. Henri Poincaré **59**, 421–444 (1993)
143. Pulé J.V., Zagrebnov V.A.: *Proof of the variational principle for a pair Hamiltonian boson model*, Rev. Math. Phys. **19**, 157–194 (2007)

144. Quagebeur J.: *A non-commutative central limit theorem for CCR-algebras*, J. Funct. Anal. **57**, 1–20 (1984)
145. Quaegebeur J., Verbeure A.: *Relaxation of the Ideal Bose Gas*, Lett. Math. Phys. **9**, 93–101 (1985)
146. Robinson D.W.: *A theorem concerning the positive metric*, Commun. Math. Phys. **1**, 80–94 (1965)
147. Robinson D.W.: *Bose-Einstein condensation with attractive boundary conditions*, Commun. Math. Phys. **50**, 53–59 (1976)
148. Roepstorff G.: *Correlation inequalities in quantum statistical mechanics and their application in the Kondo problem*, Commun. Math. Phys. **46**, 253–262 (1976)
149. Ruelle D.: *Statistical mechanics*, Benjamin, New York-Amsterdam (1969)
150. Requardt M.: *Fluctuation operators and spontaneous symmetry breaking*, J. Math. Phys. **43**, 351–372 (2002)
151. Sewell G.L.: *KMS conditions and local thermodynamic stability of quantum lattice systems II*, Commun. Math. Phys. **55**, 53–61 (1977)
152. Sewell G.L.: *Quantum theory of collective phenomena*, Oxford University Press, 1986; second edition 1989
153. Sewell G.L.: *Off-diagonal long range order and superconductive electrodynamics*, J. Math. Phys. **38**, 2053–2071 (1997)
154. Sewell G.L.: *Quantum Mechanics and Its Emergent Macrophysics*, Princeton University Press, 2002
155. Simon B.: *The Statistical Mechanics of Lattice Gases, Volume I*, Princeton University Press, Princeton, New Jersey, 1993
156. Spohn H.: *Large Scale Dynamics of interacting particles*, Springer, Berlin, 1991
157. Stern H.: *Broken symmetry, sum rules, and collective modes in many-body systems*, Phys. Rev. **147**, 94–101 (1966)
158. Straeter R.F., Wightman A.S.: *PCT, spin and statistics and all that*, W.A. Benjamin, INC, New York, Amsterdam (1964)
159. Thirring W.: *Quantum mechanics of large systems*, Springer-Verlag, New York, 1983
160. Vandevenne L., Verbeure A.F., Zagrebnov V.A.: *Equilibrium states for the Bose gas*, J. Math. Phys. **45**, 1606–1622 (2004)
161. Vanheuverzwijn P.: *Generators for quasi-free completely positive semigroups*, Ann. Institut Henri Poincaré **29**, 123–138 (1977)
162. Verbeure A.F.: *The condensate equation for non-homogeneous bosons*, Markov Proc. Related Fields **13**, 289–296 (2007)
163. Verbeure A.F.: *Long range order in lattice spin systems* J. Phys. A: Math. Theor. **42**, (2009) 232002
164. Verbeure A., Zagrebnov V.A.: *Phase transitions and algebra of fluctuation operators in an exactly soluble model of a quantum anharmonic crystal*, J. Stat. Phys. **69**, 329–359 (1992)
165. Verbeure A., Zagrebnov V.A.: *About the Luttinger model*, J. Math. Phys. **34**, 785–800 (1993)
166. Verbeure A.F., Zagrebnov V.A. *Gaussian, non-Gaussian critical fluctuations in the Curie-Weiss model*, J. Stat. Phys. **75**, 1137–1152 (1994)
167. Wagner H.: *Long-wavelength excitations and the Goldstone theorem in many particle systems with "broken symmetries"*, Z. Phys. **195**, 273–299 (1966)
168. Wreszinski W.F.: *Fluctuations in some mean field models in quantum statistics*, Helv. Phys. Acta **46**, 844–868 (1974)
169. Zagrebnov V.A., Bru J.-B.: *The Bogoliubov model of weakly imperfect Bose gas*, Phys. Rep. **350**, 291–434 (2001)

170. Ziff R.M., Uhlenbeck G.E., Kac M.: *The ideal Bose gas (revisited)*, Phys. Rep. **32(C)**, 169–248 (1977)
171. Zubarev D.N., Tserkovnikov Yu.A.: *On the theory of phase transition in non-ideal Bose gas*, Sov. Phys. Dokl. Nauk **120**, 991–994 (1958)

Index